PLANTS IN CHANGING ENVIRONMENTS

Linking physiological, population,
and community ecology

F. A. BAZZAZ

Harvard University

CAMBRIDGE
UNIVERSITY PRESS

Published by the Press Syndicate of the University of Cambridge
The Pitt Building, Trumpington Street, Cambridge CB2 1RP
40 West 20th Street, New York, NY 10011-4211, USA
10 Stamford Road, Oakleigh, Melbourne 3166, Australia

First published 1996

Printed in Great Britain at the University Press, Cambridge

A catalogue record for this book is available from the British Library

Library of Congress cataloguing in publication data
Bazzaz, F. A. (Fakhri A.)
Plants in changing environments : linking physiological, population, and
community ecology / F.A. Bazzaz
p.   cm.
Includes bibliographical references (p.   ) and index.
ISBN 0 521 39190 3 (hc). – ISBN 0 521 39843 6 (pb)
1. Plant succession.   2. Vegetation dynamics.   3. Plant ecology.
I. Title.
QK910.B39   1996
581.5′24 – dc20   96–6155 CIP

ISBN 0 521 39190 3 hardback
ISBN 0 521 39843 6 paperback

Will Wilson
Zoology Department
Duke University
Durham, NC 27708-0325

Forces of nature and human intervention lead to innumerable local, regional, and sometimes global changes in plant community patterns. Irrespective of the causes and the intensity of change, ecosystems are often naturally able to recover most of their attributes through natural succession. With the heightened interest in the fate of the biosphere, the emphasis on sustainable development worldwide and the possible consequences of global climate change, the study of succession and ecosystem recovery takes on added urgency. Successional theory will play a major role in ecosystem preservation, management, rehabilitation, and restoration.

Fakhri Bazzaz takes a broad view of disturbance and recovery, from filling of small gaps to the revegetation after clearing of large areas for agriculture and forestry. The book integrates and synthesizes information on how disturbance changes the environment, how species function, coexist, and share or compete for resources in populations and communities, and how species replace each other over successional time. Furthermore, the book shows how a diverse array of plant species from different successional positions have been used to examine fundamental questions in plant ecology by integrating physiological, population, and community ecology. The basic philosophy of the work is that the physiological activities of individuals and the ecology of populations do not happen in a vacuum. Individuals in a population are imbedded in a community matrix, and are influenced by the presence and activities of other individuals and populations of the same and of other trophic levels. Furthermore, physiological and population processes strongly influence community composition and dynamics. This complexity makes the study of ecosystem recovery at once difficult, challenging, and exciting.

Graduate students and research workers in plant ecology, global change, conservation, and restoration will find the perspective and analysis offered by this book an exciting contribution to the development of our understanding of plant successional change.

PLANTS IN CHANGING ENVIRONMENTS

Contents

Acknowledgments

I thank my current and former students and post docs. who greatly contributed to the work presented in this book. They freely shared their ideas with me and with each other, and by their critical thinking, insightful discussions, cheerful attitudes, and comradery made our lab., for me, (and I hope for them) a very exciting place.

Over my academic career, I have been fortunate to have had advice, support, inspiration and help from many teachers, colleagues, students and post-doctoral associates, and lab. visitors too numerous to mention by name. In particular, I would like to especially thank L. C. Bliss, W. R. Boggess, J. S. Boyer, J. L. Harper, P. J. Grubb, H. A. Mooney, R. M. May, P. W. Price, P. M. Vitousek, F. I. Woodward, G. M. Woodwell, D. A. Levin, E. H. Franz, A. R. Zangerl, R. W. Carlson, D. J. Raynal, S. T. A. Pickett, J. A. D. Parrish, T. D. Lee, J. S. Coleman, E. Reekie, K. Garbutt, B. Schmid, D. C. Hartnett, K. D. M. McConnaughay, P. Wayne, and E. D. Fajer.

Several members of our lab. also critically read and commented on various chapters of this book. In particular, I thank Sonia Sultan, Susan Bassow, Sean Thomas, David Ackerly, Timothy Sipe, and Gary Carlton.

Kristen Norweg, with the help of Jennifer Cox and Charlotte Kaiser cheerfully corrected my word processing of the manuscript several times over. She and Glenn Berntson used their impressive computer skills to draw the figures. Beth Farnsworth drew some of the sketches, skillfully edited the entire book and offered many helpful suggestions. Michal Jasienski greatly helped with the bibliography.

Finally, I thank my wife and our daughter and son, for their endurance, support, and love.

1
Introduction and background

Succession, ecosystem recovery, and global change

Forces of nature lead to innumerable local, regional, and sometimes global changes in plant community patterns. Even early humans undoubtedly observed vegetational change and may have used that change to their advantage. They saw the power of natural forces to change vegetation and used them, especially fire, to 'manage' ecosystems for their own needs. The exchange of materials between the atmosphere, the biosphere, and the ocean was largely controlled by natural events in the past but is currently being greatly modified through increased human activities (Fig. 1.1). Irrespective of the causes and the intensity of change, ecosystems are often naturally able to recover most of their attributes through natural succession. They can also be repaired through human intervention such as land reclamation. Because of the recent enormous rise in the human population and its per capita consumption of natural resources (Fig. 1.2), the earth is becoming increasingly occupied by successional ecosystems in various stages of recovery. With the recent heightened interest in the fate of the biosphere (e.g. Lubchenco *et al.* 1991, Woodwell and MacKenzie 1995), the emphasis on sustainable development worldwide and concern about the possible consequences of global climate change (Houghton *et al.* 1990, 1992, Vitousek 1994), the study of succession and ecosystem recovery takes on an added urgency. Successional theory will play a major role in ecosystem preservation, management, rehabilitation, and restoration.

Broadly viewed, succession is the recovery process of vegetation following any disturbance. Successional studies have focused on repeatedly observable transitions in community composition following either the exposure of new substratum (usually referred to as primary succession) or the disruption of existing communities by various agents of disturbance (secondary succession).

1

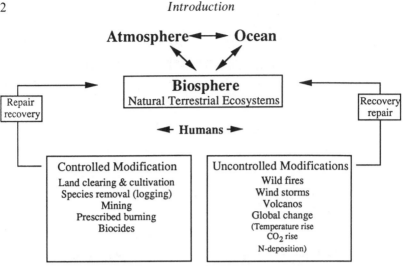

Fig. 1.1. The impacts of humans on atmosphere–biosphere–ocean interactions. Disturbance and recovery through succession.

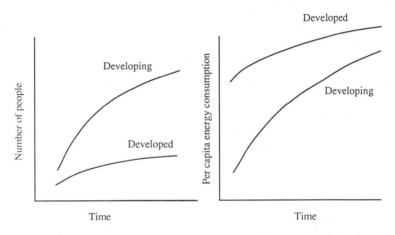

Fig. 1.2. Projected increase in human population and its impact on the environment through consumption of natural resources, in 'developed' and 'developing' countries.

Classically, succession was viewed as a subset of many different types of ecosystem recovery, from new colonizers filling in small vegetation gaps to regeneration during very large climatic changes such as deglaciation or global warming.

Historical perspective

We do not know if scholars of older cultures in China, the Middle East, and the Americas formally considered community change, but we do know that Western philosophers and naturalists from Theoprastus to Thoreau thought and wrote about vegetational change. The study of plant community dynamics in Western science seems to have first appeared in the formal sense in the nineteenth century, in A. Kerner's description of the vegetation change in the Danube River Basin. In the United States, H. Cowles pioneered the study of 'primary succession' on newly exposed habitats with his classic work on zonation of vegetation of the sand dunes of Lake Michigan. However, F. E. Clements is better remembered as the father of successional theory. Clements derived much of his intellectual framework form German phytogeography, and his research experiences were restricted largely to the stable communities of the American midwest. His views were also influenced by the then predominant paradigm in geology: peneplaination, or the long term erosion of mountains to plains. As a result, he viewed plant communities as complex quasi-organisms and succession as the progressive development of the quasi-organism to a predetermined climax.

With only a few exceptions (e.g. H. A. Gleason), early ecologists considered succession a community-level phenomenon. They sought to identify the 'stages' of succession, the sequence of their appearance, the duration of their persistence, and their orderly progression to uniform and predictable endpoints or 'climaxes.' In contrast, more recent views of succession claim that succession is a multidirectional, probabilistic process which can have more than one endpoint (review in Pickett *et al.* 1987). Views about community organization, while not clearly allied with succession, are inextricably linked to successional theory. The apparent disjunction of community organization theory from successional theory stems from the fact that the former has been the domain of animal ecologists and the latter the domain of plant ecologists, and for a long time the two have been developing independently of each other. The debate over the Clementsian and Gleasonian views about the nature of the plant community is legendary (see McIntosh 1980) and it continues unabated today. There is probably no subject in plant ecology that has been so extensively treated. Unfortunately, many plant ecologists have spent much of their energy and time debating these views about communities, and have rarely paid the appropriate attention to the scale at which the issues are being considered (see Allen and Starr 1982, O'Neill *et al.* 1986).

Scale

Appropriate consideration of scale clarifies much of the debate and resolves what may appear to be irreconcilable views. That there are physiognomically predictable communities on a regional scale cannot be denied. For example, coniferous forests dominate a huge area in the United States and Canada and stretch across Europe and much of Siberia, and temperate deciduous forest dominates most of the eastern United States. However, patchiness can be detected at several finer scales within these regions. Various combinations of species are present in each patch and different successional trajectories are possible. Moreover, individuals in a stand do not behave independently from their neighbors; they interact with each other through the shared or contested resources of the habitat. They are part of each other's environment. Although plant species have distinct genetic make-ups that influence their physiologies and distributions (Gleason 1926), individuals do interact with neighbors and can therefore modify each other's activities and distributions. Because of overlap in requirements for growth and reproduction, species can occur together in combinations as recognizable communities (Weaver and Clements 1938). Confusion between the 'stand' (which in plant ecology is a tangible unit of the landscape that can be measured and manipulated) and the 'community' (which is an abstraction obtained from observing and measuring several stands; Oosting 1956) has caused additional confusion in successional theory and plant community ecology. A critical appreciation of the importance of scale can clarify some fuzzy concepts in plant community ecology. This clarification can go a long way in successfully embedding successional studies into functional ecology, which could explain extant communities, their past, and their evolutionary pathways.

The consideration of successional processes at the ecosystem level, initially motivated by the writings of Ramon Margalef, began with the publication of E. P. Odum's paper, 'Strategy of Ecosystem Development' (1969). Based on studies of succession in oldfields in the southeastern United States, Odum proposed a list of attributes of successional populations and ecosystems. This seminal paper, more than any other, has stimulated and motivated much of the research in ecosystem dynamics. Disturbance ecology and the study of ecosystem recovery during succession have moved to positions of prominence, particularly with the publication of the work of F. H. Bormann, G. E. Likens and their associates on the Hubbard Brook watershed in the northeastern United States (see Bormann and Likens 1979). Controls on primary productivity, nutrient dynamics, and transport across system boundaries have become subjects of study in many systems worldwide. The former International Biological Program (IBP) and the more recent Long Term Ecological Research (LTER) network are integrated activities for, and large scale expressions of interest in, understanding ecosystem dynamics.

Because the plant's environment is ever-changing on a number of time scales, there is little in the ecology of plants that is unrelated to succession in the broad sense. Therefore, successional theory can form the basis for much of the life history evolution, population dynamics, competitive interactions, nutrient dynamics, and community organization of plants. Practical issues in ecology, such as restoration of damaged ecosystems and predicting consequences of global change, also draw heavily on theories of succession. With the now firm understanding that all communities are dynamic and patchy at several scales, there should be no distinction between plant community dynamics and successional studies. Ecosystem dynamics and successional change are best viewed as one and the same (Miles 1979). In fact, the distinction between community and ecosystem is itself misleading. 'A community does not become an ecosystem by adding the abiotic environment; it is an ecosystem because it is inseparable from the abiotic environment' (Bazzaz and Sipe 1987). While community ecology emphasizes the producers, no community exists in nature without the critical involvement of other trophic levels. It stands to reason, then, that whatever influences successional rates and trajectories also influences ecosystem structure and function. A mechanistic understanding of the fundamental and generalizable principles of succession, and their applications to global change questions, requires great depth and breadth in several areas of ecology. Knowledge of some aspects of molecular biology, plant physiology, population and community (ecosystems) ecology, and mathematical modeling is necessary to satisfactorily answer questions about succession and climate change. The ability to communicate in the language of the physical scientists is becoming more and more important for the profitable exchange with the physical scientists of information needed for assessment of the impact of global change.

A list of important works on succession can include hundreds of citations. Interest in succession has become even more prominent under the new heading 'disturbance' (reviews in White 1979, Mooney and Godron 1983, Pickett and White 1985). Extensive treatments of succession by Clements (1916), Loucks (1970), Drury and Nisbet (1973), Horn (1976), Pickett (1976), Connell and Slatyer (1977), Hayashi (1977), Gorham *et al.* (1979), Miles (1979), McIntosh (1980), Noble and Slatyer (1980), Peet and

Christensen (1980), Van Hulst (1980), Bornkamm (1984), Finegan (1984), Shugart (1984), Huston and Smith (1987), Pickett *et al.* (1987), Numata (1990), and Osbornová *et al.* (1990) have been particularly illuminating. These reviews present extensive information on the history, philosophy, classification, causes, and mechanisms of succession. Amazingly, despite this great volume of work on the causes and mechanisms of succession, the phenomenon is still without a general theory (see Huston and Smith 1987, Huston 1994). The development of such a theory (assuming there can be one) will help to address some of the emerging and relevant issues of global change and their consequences for human welfare, sustainable development, and the preservation of biological diversity. The recent applications to this field of chaos theory, fractal geometry, vector calculus, and the strong emergence of the new science of complexity can profoundly impact such theories of succession. Needless to say, exciting and challenging times lie ahead for new generations of ecologists who will be armed with these new skills to solve fundamental questions in ecology.

The scope of this book

This book does not focus on the details of successional history and the progress toward the development of classical successional theory. These aspects have been thoroughly addressed before by several authors (see Pickett *et al.* 1987). Instead, this book considers the broader view of disturbance and recovery, from filling of small gaps to the revegetation after clearing of large areas for agriculture and forestry. The book integrates and synthesizes information on how disturbance changes the environment, how species function, coexist, and share or compete for resources in populations and communities, and how species replace each other over successional time. Furthermore, the book shows how we used a diverse array of plant species from different successional positions to examine fundamental questions in plant ecology by integrating physiological, population, and community processes. The basic philosophy of our work is that the physiological activities of individuals and the ecology of populations do not happen in a vacuum. Individuals in a population are embedded in a community matrix, and are influenced by the presence and activities of other individuals and populations of the same and of other trophic levels. Furthermore, there is a reciprocal effect, in that physiological and population processes strongly influence community composition and dynamics. This complexity makes the study of ecosystem recovery at once difficult, challenging, and exciting. It also dictates simultaneous work in microenvironmental measurements, physiological responses to these changes, feedback loops, and population and community ecology.

This book draws heavily on work carried out by me and my associates first at Illinois and later at Harvard in oldfields, temperate forests, and tropical rainforests. But it is not only a monograph; this work is integrated with the relevant literature, although it is not intended to be a comprehensive review of the literature. The book does not present a 'special case,' but uses our findings from these diverse ecosystems to develop a broad understanding of species' responses to rapidly changing environments, especially during vegetation recovery after disturbance. In our studies and throughout the book, we emphasize that even small disturbance events that initiate succession can modify the levels of resource availability in a profound way and can greatly influence many interconnected responses of plants to these changes, including negative and positive feedbacks at the individual, population and community level (Fig. 1.3).

Work in the Illinois fields

I started my work on plants of successional habitats with the aim of finding out as precisely as I could why certain species replaced each other, above and beyond what Clements and others already knew: that replacement had something to do with availability of propagules, differences in life cycles, competitive superiority, and site modification. I studied secondary successional stands in the hills of southern Illinois, where many fields had been abandoned because of poor productivity. The fields are located on thin, highly eroded, acidic soils and are thus deficient in nutrients. H. A. Gleason had done his Master's thesis work in the same area decades before me. At that time, the prevailing model for studying vegetation recovery was oldfield succession after forest clearing, agricultural use and subsequent abandonment. Research on oldfield succession was energized through the pioneering work of H. J. Oosting and his students, in fields of North Carolina and elsewhere that were abandoned during the Depression. Like several authors before me (e.g. Drew 1942, Oosting 1942, Bard 1952, Quarterman 1957), I wanted first to establish the patterns of change over time by sampling the vegetation of several fields known to differ in the approximate length of time since last cultivation. In so doing, I was substituting space for time. Having been educated under the influence of Clementsian thinking, I was hoping to find a uniform domination in each age group of fields by a few species, which could then be ranked by some measure of their

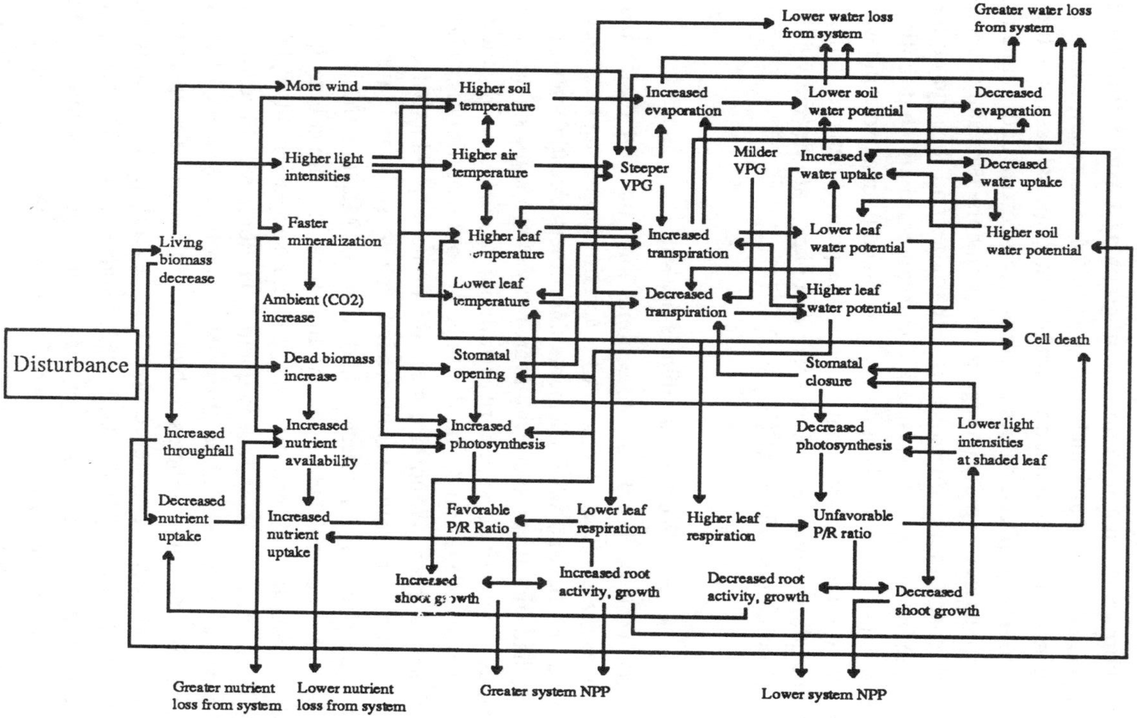

Fig. 1.3. The great complexity of physiological responses and their many feedbacks caused by a single disturbance event (modified from Bazzaz and Sipe 1987).

relative importance, and from which a generalized community pattern of succession would emerge.

My first surprise and a cause of some frustration and despair was that even in small fields of a few hectares that had been abandoned only recently, and which were apparently treated fairly uniformly prior to abandonment, there was much variation in vegetation. Adjacent patches with different dominants were quite common. Patches in older fields appeared to have vegetation typical of younger fields. Invasion of late-successional trees such as *Quercus* occurred only in middle-aged, clonally-spreading, patchily distributed thickets of *Sassafras* and *Diospyros* (Bazzaz 1968). In many fields, erosion created a soil-depth gradient on which different species occurred in different locations and with different breadths of occupation, some narrow and others quite broad (Fig. 1.4) (Bazzaz 1969). It became clear that certain of these species were good indicators of specific habitat factors throughout a large geographic region. The emerging notion of species diversity, a simple yet powerful central construct in community ecology, became rather complicated. Instead of the expected monotonic increase in species diversity over time, patterns were complex and difficult to scale in these patchy fields (Bazzaz 1975). With time and more experience, I realized that the Clementsian notions of uniformity, directionality, reproducibility, and the idea that succession results in uniform regional climaxes are uncommon and have only limited explanatory power in explaining vegetational change. That convinced me to go beyond description and to ask questions about where, when, and why species appear in succession. How are these patches created? What is their environment? What are their fates? Why and how are patches replaced by others? The scale at which field observations should be made and attributes expressed became major puzzles for me.

Beyond the Clementsian generalizations about causes of succession – 'denudation,' 'dispersal,' 'establishment and reproduction,' and 'site preparation' by one group of species for the next group to invade – there was limited information about the actual mechanisms of succession. The works of Catherine Keever (1950) on early stages and of F. Herbert Bormann (1953) on middle stages of temperate forest succession stood out as exceptions. Working from this foundation, I asked how different physiological characteristics and life history features of the species involved influence successional replacement. I only scratched the surface of this vast question in my Ph.D. dissertation, but I continued to explore it for several years afterward. My initial emphasis was on the controls of

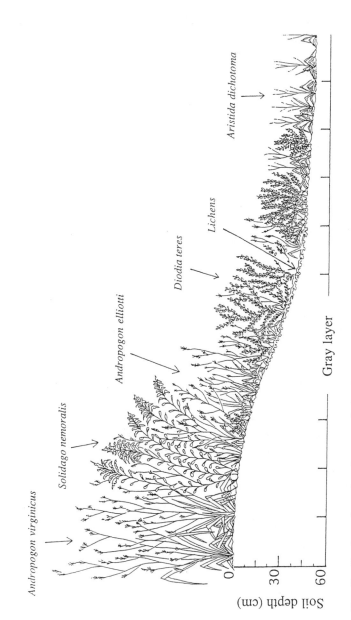

Fig. 1.4. The distribution of species along a soil depth gradient created by different degrees of erosion in a mid-successional field.

germination, photosynthesis, growth, and other physiological attributes, in relation to the physical environment along successional gradients. I wanted to discover how species respond to the physical environment and how this response may lead to different paths of succession. At first, I did not give numbers of individuals the attention they deserved. They were expressed as density (number of stems per unit area), usually lumped together with frequency (number of sampling plots in which a given species is present) to produce importance values (IV) for dominance ranking of species in the community. At that time, demography and other aspects of population biology were practically unknown to plant ecologists. They had not yet been brought to center stage through the work of John Harper and his students (e.g. Harper 1967, 1969, 1977, 1985, White 1985). Soon after, however, our work involved the integration of physiological, population, and community aspects of succession. As will be demonstrated in the following chapters, we believe that the separation between these three branches of plant ecology is artificial, and that an understanding of processes like succession requires this kind of integration.

Like my early work, much research on oldfield succession was carried out on inherently unproductive sites with poor soils. These were the lands that were commonly available for study. However, because of heavy fertilizer application during cultivation, some of these abandoned fields were initially high in nutrients, but quickly became depauperate through leaching. Patterns of primary productivity in succession usually follow those of soil fertility, as has been demonstrated by E. P. Odum (1960) in the South Carolina fields. However, fields in different locations do differ in their soil type, their innate fertility levels, and the amounts and kinds of prior fertilizer application. This site variability has resulted in contradictory conclusions from various studies of succession. Availability of soil resources, especially nutrients, critically influence individual plant growth, population dynamics, competitive interactions, community structure, and successional change. There is little doubt, therefore, that these differences in results can be explained and understood if soil resource availability (nutrients and moisture) are appropriately scaled. Interestingly, Clements clearly recognized that initial soil fertility influences the rate of succession, and yet this did not alter his comunity-as-organism paradigm.

Most studies of succession in fields substituted space for time (*sensu* Pickett 1989); that is, fields abandoned for a known number of years were simultaneously studied to develop a successional series. In 1969, I initiated a long-term study involving six treatments in a relatively uniform field that was under continuous cultivation for 62 years prior to the start of the experiment. The treatments were three soil disturbance times on two strongly different soil types. The aim of the study was to test how the initial vegetation might differ among the treatments and whether these adjacent treatments converged or diverged with succession – a test of the influence of initial conditions on community structure. We examined recruitment patterns, community structure, patterns of species diversity, changes in plant populations, biomass accumulation, and the physiology of colonization.

My thinking and approaches to the study of successional change have undergone much evolution over the last two and a half decades. During this time, several students have worked with me, and our accumulated knowledge of plants in successional environments has guided our research. Frequently, our emphasis has shifted from the central questions about succession to other exciting issues in physiological, population, and community ecology, especially those that link levels of organization. In these studies, plants from various successional stages were used as study organisms to answer fundamental questions about the ecology of plants in general. We used as models species in the genera *Erigeron, Ambrosia, Abutilon, Polygonum* (C_3 annuals); *Amaranthus, Setaria* (C_4 annuals); *Aster, Solidago* (clonal); and *Populus, Prunus, Betula, Quercus, Acer* (perennials) in a variety of physiological and demographic studies. Among tropical successional plants, we used *Piper, Cecropia, Ficus, Heliocarpus,* and *Shorea* to study plasticity, acclimation, and species coexistence. Several of the early-successional annuals and deciduous forest seedlings were used both as individuals and in 'mesocosms' (Lawton 1995) to elucidate the effects of global change on plant competitive interactions and community structure (reviews in Bazzaz 1990*b*, Bazzaz and Fajer 1992, Bazzaz and McConnaughay 1992, Bazzaz *et al.* 1996). We studied patterns of resource use, acclimation and plasticity, reproductive allocation and costs, clonal integration, plant demography, plant–plant interactions, the structure of the plant niche, and habitat selection. We placed special emphasis on linking various aspects of plant ecology in order to gain a cohesive understanding of the behavior of plants as groups of interacting individuals in populations and communities in their ever-changing natural environments. It is important to remember that much of this book is about these topics; hence the title 'Plants in Changing Environments: Linking Physiological, Population, and Community Ecology.'

Interest in CO_2 as a plant resource and as a critical element of global environmental change has dramatically increased, with the recognition by ecologists that the 'Keeling CO_2 curve,' which shows increasing levels of CO_2 in the atmosphere and much seasonal variation, could have profound implications for plant growth and productivity and is also highly influenced by them. If species of the same community respond very differently to elevated CO_2 and other elements of global change, then competitive hierarchies could change and the process of community organization and ecosystem recovery after disturbance could also change (Bazzaz *et al.* 1985*a*). Over the past two decades, extensive work on the response of plants to elevated CO_2 (Strain 1987) and other environmental factors such as nutrients, soil moisture, light, temperature, and air pollutants has prodded us to consider disturbance and recovery at very large (regional and global) scales. Our work currently has closer and more fruitful contact with ecosystem scientists, remote sensing experts, modelers, and atmospheric chemists.

Forest regeneration

Over the last decade, we have focused on the process of regeneration in forests with the aim of linking oldfield succession to forest succession. We have carried out extensive studies on changes in the microenvironment brought about by the clumping of invading shrubs and trees in oldfields and forest gaps, the physiological and demographic characteristics of species that invade these disturbed sites, dynamics of species recruitment and replacement, and how these species might partition the gap environment and therefore coexist. Our studies are being extended to investigate ecosystem-level exchange of carbon dioxide and water between the forest and the atmosphere in order to permit scaling from leaf-level measurement to estimates of whole-forest flux. By using plants from both temperate and tropical successions, we hope to develop a mechanistic understanding of succession and ecosystem recovery in general, and in the process, elucidate some principles of the biology of plants.

Because of this particular approach, the patterns of funding, and the diverse interests and expertise of my associates over the years, our work on understanding the behavior of successional plants has not always proceeded as a united front. Indeed, some parts of the puzzle were not added in chronological order and others still remain unknown. To use two words I first heard from John Harper, our progress was made through a mixture of 'phalanx' and 'guerrilla' tactics. In telling our story in this book, I will try to remedy this in a *post hoc* fashion by presenting various topics in a more logical, rather than chronological, order.

2
Plant strategies, models, and successional change: a resource-response perspective

Strategies and models

Physiological ecology and population ecology have remained separate for a long time. Traditional approaches in physiological ecology put strong emphasis on the mechanisms of plant adaptation to the environment. Similarity among individuals is sought. An individual or a few individuals are assumed to represent the 'species.' Mean response is more important than the variance of response. Increased availability of resources is customarily inferred from improved performance, such as increased per unit leaf photosynthetic rate or higher growth. Environmental variability is assessed at the level of the module (e.g. single leaf) or of the whole plant. Observations and experiments can capture a good part of plant development. In contrast, population ecology emphasizes difference among individuals, since natural selection operates on this variability. Variance is more important than the mean. Increased resource availability may lead to earlier competition and a decrease in population size. Environmental variability is considered relevant, both at the scale of the individual and at the scale of pollen and seed dispersal of the members of the population. In populations it is important to recognize that observations and experiments on plants are necessarily made through a very narrow window of time in their history, and they may capture only a glimpse of that history. Populations have a long past. The evolution of observed traits has resulted from the interaction of phenotypic variation and selection in the past, and the population's current variation and selection are critical to its future. Understanding the behavior of plants in successional habitats requires investigating strategies and models of both individuals and populations. To understand plant strategies and successional change, therefore, we used both physiological and demographic approaches because responses to resources occur at both physiological and population levels.

In this book the term 'plant strategies' is taken to mean evolutionarily developed (genetically based) patterns of response to the elements of the environment that are likely to be encountered in the plant's habitat. The term is commonly used at the species level to group plants into 'ecological groups,' 'functional types,' or 'guilds.' (Root 1967, Whittaker 1975, Smith *et al.* 1995). Although species (and individuals) are expected to respond differently to their environment, grouping is essential to derive generalizable principles of succession and ecosystem recovery. The strong desire (and need) to make predictions about the impact of global change on a large scale has heightened this interest. The diversity of life on our planet is too large to know how each individual species may respond to various elements of global change.

Several attempts have been made to identify life history trends in plants and fit them into the general theories about life history evolution developed for organisms in general. A prominent example is the r-K continuum life history of strategies proposed by MacArthur and Wilson (1967). Adapting their definitions to higher plants, *r-strategists* are plants that occupy disturbed habitats and are therefore unable to persist in a given location for a long time (Solbrig 1971). They are usually short-lived, their populations grow rapidly, and their populations are regulated by density. In contrast, *K-strategists* persist in the same habitat for a long time relative to their life span, and at numbers hovering near the carrying capacity of that habitat for the particular species. Their population regulation is also strongly density-dependent. Using this sort of classification, plants of early-successional habitats are generally considered r-strategists, and later-successional plants are considered K-strategists (Odum 1969). In natural plant communities, however, particularly at the landscape scale, disturbance is common, and therefore successional change is so prevalent that no population will exist in the same habitat for a long time. Rather, species replace each other, and competitive interactions do not commonly lead to the long-term dominance of any one species. Because of this, and due to the fact that plants are modular in structure and rather plastic, the designation of r- and K-strategies is of only limited explanatory value in plant ecology.

The three primary strategies in plants proposed by Grime (1977) – competitive, stress-tolerant, and ruderal (CSR) – allow any species to have a combination of strategies in the evolution of its life history. Grime considers these primary strategies the tips of a triangle and believes that any species can be located within the triangle according to its life history

Resource-response perspective

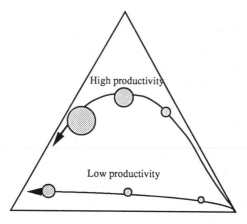

Fig. 2.1. Successional trajectories in relation to competitive, stress-tolerant, ruderal (CSR) strategies (modified from Grime 1979).

features, which determine its proximity to the corners of the triangle. For example, Grime (1979) shows that since succession in a clearing in temperate forests on a moderately fertile site is characterized by a quick development of plant biomass and a rapid replacement of the early-successional species by fast-growing shrubs and trees, the succession trajectory rises towards the 'C' strategy. At later stages, the course of succession begins to deflect downwards toward the stress tolerance corner of the triangle (S), reflecting the decreasing importance of species with high rates of resource capture and loss, and the increasing importance of species that retain resources, especially nutrients, in the plant tissue (Fig. 2.1). Secondary succession on sites with low soil fertility has a more shallow parabola, and plant biomass is smaller because of the earlier onset of nutrient limitation. Succession on different sites therefore proceeds from right to left with differing proximity to the apex of the triangle. While ecologists have often made use of the CSR framework, they have not applied this graphical approach widely to successional studies. Thus, it remains to be seen if great insights into mechanisms of succession can be gained from this sort of analysis.

Tilman (1982, 1986) explains succession in terms of resource ratios, especially between nitrogen and light. He assumes that an area changes with succession from a low-nitrogen and high-light environment to high-nitrogen and low-light environment. His model fits many primary successions on sandy soil and newly exposed sites, as well as some secondary successions that occur on nutrient-poor soils. He considers the ultimate indicator of success to be not transient dominance, but persistence in the site. In this model it is assumed that competing species experience trade-offs in their resource requirements such that a superior competitor for one resource is an inferior competitor for other resources. Thus, early-successional plants in Grime's model are efficient resource capturers, and late-successional plants in Tilman's model are those with greater 'staying power' at the population level. Evidence from many physiological and community studies suggests that early- and late-successional plants, in general, can be described by Grime's and Tilman's respective views. Despite much discussion about the differences between Tilman's and Grime's views about community organization (see Goldberg 1990), in terms of succession these two views appear complementary rather than contradictory (Grace 1990).

Despite the serious shortcomings in fitting plant populations to the r-K related Lotka-Volterra formulations, models of succession based on the Lotka-Volterra population growth equation have been used to simulate succession (Huston and Smith 1987). With the appropriate adjustment to express r, N, and K in terms of biomass, rather than numbers of individuals, as is called for in the original logistic equations, and by expressing the competitive coefficient α_{ij} in terms of the effect of relative size of a neighbor (j) on the target organism (i), the simulations produce succession-like behavior. While all of these modeling approaches are promising in that they do produce reasonable predictions, there is still a long way to go before succession is closely simulated and its long-term trajectories accurately predicted.

Huston and Smith (1987) explain succession on the basis of competitive interactions among individuals, emphasizing inverse correlations among life history traits in plants. They consider previous models of succession to be descriptive rather than mechanistic, and propose a simulation model based on the JABOWA/FORET forest simulation model (Shugart 1984). In this model, they assume quite correctly that interactions occur among individuals rather than among populations. They show that various combinations of selected life history traits and physiological attributes in the model can reproduce population dynamics found in natural successions (Fig. 2.2). In the model, the birth, growth, and death of each individual is followed through time, each individual is assigned species-specific life history traits (maximum size, age, growth rate, and shade tolerance), and the availability of light is modeled explicitly. Unlike Tilman's model, which emphasizes populations, these authors consider interactions among individuals *and* assume non-equilibrium dynamics. Other recent extensions

Resource-response perspective

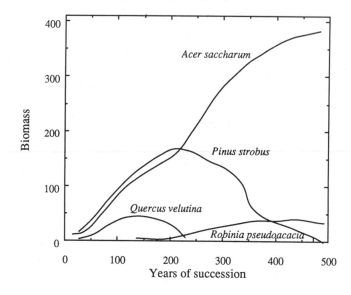

Fig. 2.2. A simulation, using a modification of the logistic equation, of forest succession after clear cutting (modified from Huston and Smith 1987).

of the JABOWA-FORET models also include substantial modification. For example the FORSKA model (Leemans 1991, Prentice *et al.* 1993) contains growth and mortality submodels that are substantially different from the original model. ZELIG (Smith and Urban 1988) includes spatial dimensions as well.

Recently, Pacala and his associates (Pacala *et al.* 1993) developed a model (SORTIE) that, unlike the JABOWA-FORET, assigns the response of a species to one of three shade tolerance classes and does not differentiate among species in a community in their growth-dependent mortality. SORTIE is spatial and mechanistic. It contains species-specific functions to predict each species' dispersal, establishment, growth, mortality, and fecundity. In the model the performance of each individual influences its neighbor by depleting resources. The model uses measurable data from the field to understand community structure and dynamics and has been used to predict forest succession in the northeastern United States.

Despite much progress in modeling succession, the tension between simplicity of models for prediction and complexity to include all relevant parameters remains unresolved. Because errors in models can interact non-additively (Rastetter *et al.* 1992) model complexity may overwhelm biological reality.

The equilibrium paradigm has dominated thinking about plants' responses to their environment despite much evidence that it plays no significant role in these processes. Undoubtedly, the mathematical simplicity of the logistic growth of individuals and the Lotka-Volterra-type population growth models have encouraged and sustained such thinking. Observations on the appropriate temporal and spatial scales would show that equilibria are usually transient and are quickly disrupted by disturbance events. Nevertheless, current arguments about resource capture and allocation between roots and shoots are influenced by implicit assumptions about equilibrium conditions and a high degree of permanence of vegetation. Individual plants may change greatly during their lives and plants adopt different strategies throughout ontogeny: a 'ruderal' seedling may be a stress-tolerant sapling. Disadvantaged juveniles in the understory can later become dominant individuals that reproduce, disperse their propagules, and die. And, despite much flexibility in response, the acquisition of resources above- and belowground must ultimately be balanced. Nitrogen and other elements taken up by roots from the soil are needed for carbon gain by shoots from the air, and carbon is required for nitrogen uptake – a natural loop broken only for purposes of study. Switches in allocation of photosynthate between shoots and roots can occur within minutes when the balance is abruptly disturbed (e.g. by herbivore consumption). Rapid shifts in allocation of resources to various functions can be advantageous in environments in which resource supply can rapidly change. The use of isotopes, particularly ^{11}C and ^{15}N, should help in discovering the rate and magnitudes of these shifts under specific changes in resource supply and plant tissues removals, e.g. by herbivores. It appears now that the grip of equilibrium dynamics on thinking about community organization is easing and that many ecologists view communities as being in non-equilibrium states most of the time.

Resources, controllers, and signals

Most disturbances do not completely destroy the vegetation of a site. Disturbances can make more resources available for enhanced growth of those individuals that survive, and for the new recruits from both within and outside the site. Plant species of the remnant community, as well as new recruits, are likely to differ in their ability to *capture* resources from their environments. Disturbance also changes the patterns of resource availability, which may simultaneously influence many activities of the plants, thus leading to noticeable changes in their behavior. Of course, unless growth is closely related to reproductive strategies, the information gained by studying growth alone will be insufficient to understand the evolution of species' behaviors. The common approach in physiological ecology, to study juveniles of long-lived plants with delayed onset of reproduction to make predictions about the response of adults, suffers from this situation. It is critical to appreciate that both the *quantity* of resources that become available after disturbance, and the *timing* of that availability, are important components of the plant's environment and of the evolution of the plant's strategies in the way it responds to patterns of resources.

Environmental factors that influence plant performance can be classified as resources or controllers. Resources are consumable or depletable substances, such as nutrients, water, and light, that are required by plants for maintenance, growth, and reproduction (see Ricklefs 1990). Controllers are non-consumable factors such as temperature, that influence plant performance ('conditions' of Begon *et al.* 1990). Some environmental factors can act as both resources and controllers. For example, part of the light energy falling on a leaf is consumed in photosynthesis and part of it heats the leaf, thus influencing its activities, including the rate of photosynthesis. At the same time, certain wavelengths, such as blue light, or ratios, such as red/far-red, influence stomatal activity and extension growth. From the individual plant's view, environmental resources (and controllers) may be indistinguishable. An inherently nutrient-poor soil patch, the presence of effectively competing neighbors, or low soil moisture may all have a similar functional effect, since the plant 'perceives' all of these situations as low nutrient availability and may respond physiologically and allocationally to this general perception rather than to the specific cause.

The major resources and conditions necessary for plant growth and reproduction are easily identifiable and can now be measured accurately using sophisticated instruments (Pearcy *et al.* 1989). However, it may be more difficult to assess actual resource availability and the consequences of that availability to plants, especially at the population or community level. Soil moisture and nitrogen content in a given location may tell us very little about the availability of these resources to neighboring individuals competing for them. For example, the size of competing individuals, the placement of their roots in the soil profile, and their relative absorption efficiency can cause differential uptake rates from a common pool. Knowing how the photosynthetic rate of a leaf responds to irradiance may tell us little about the carbon gain of the individual unless we also know something about the plant's total leaf area, its display relative to incident radiation, whole-plant respiration (Mooney 1972), and the quantities of exudates released into the soil. One of the greatest challenges in functional plant ecology is to reconcile environmental fluxes and their variability as defined and measured by the investigator with the fluxes and their variability as detected and responded to by the plant. To further complicate matters, plant perception can be greatly influenced by the degree of phenotypic plasticity of the individual plant (as will be discussed in Chapter 10). Because plants can independently allocate various resources to different parts and functions, the currency of allocation must be determined in order to calculate the cost of the particular structure or function. What then is the appropriate currency of allocation, especially since the plant may be limited by different resources during its life? Carbon has been proposed as the appropriate currency since it is the investment of carbon that makes possible the acquisition of all other plant resources (Bazzaz and Reekie 1985). For example, plants allocate carbon to grow more roots to have access to more nutrients and water from their environment. In this case, all carbon acquired in photosynthesis and lost in respiration provides a precise measure of the plant's carbon economy.

Resource congruence, capture, and capacitance

It is axiomatic to say that the environment of a plant is variable in time and space. There is every reason to believe that this variability has played an important role in the evolution of plant life histories in most habitats. In many instances, spatial variation (patchiness in resource availability) itself is variable over time. A stationary individual will experience much variation during its life. For example, in early-successional fields, a wet patch could dry up over time to become a mesic patch and then a dry patch. Nutrients can become more or less available, depending on the change in soil moisture, even when nutrient quantities remain unchanged. If nitrates are lost due to the action of nutrifiers during inundation in a wet patch, then as the wet patch dries out and changes to a mesic one, it may become more suitable for plant growth in terms of soil moisture but less suitable with regard to nitrate availability – an *incongruence* between two critical resources in the same location. The spatial and temporal variability can occur in small or in large patches and may occur at different scales for various resources. Therefore, resource availability is scale dependent (Kolasa and Rollo 1991). Patterns of spatial heterogeneity in resource availability following disturbances can be described at the landscape, community, individual seedling, or cellular level, and temporal heterogeneity can be described over seconds, hours, years, or centuries.

Required resources do not necessarily become simultaneously available in optimal quantities to the plant. Resource '*congruence*' is a critical component of the environment of plants, and its patterns in various habitats have influenced the evolution of plant strategies. Congruence refers to the simultaneous availability in time and space of the major resources required by a plant. In many habitats, midday depression of photosynthesis under some field conditions results from incongruency of light and moisture availability at the leaf level. Light intensity may be sufficiently high during the time when moisture is insufficient or CO_2 concentrations are low. Early in the growing season in temperate forests, light and temperature conditions may be optimal for photosynthesis, while the availability of soil moisture and nutrients are less than optimal because of cold soils. Resource '*capacitance*,' on the other hand, is the ability of a plant to acquire resources when they are available in good supply and to keep them until other resources required for growth and reproduction also become available. Capacitance is made necessary by the fact that resources are not congruent. Capacitance is an effective strategy, especially in habitats where resource supply is unpredictable. For example, it is known that many grass species take up much of their required nitrogen during the early phases of their growth, similar to 'hoarding' by animals. This activity may confer benefit by accumulating resources in excess of immediate need, and preventing competitors from getting them. Thus, resource *capture*, *congruence*, and *capacitance* are critical aspects of plant–environment interactions. They all must be understood for a clearer appreciation of the plant–environment relationship.

Effect and response: reciprocity and asymmetry

The availability of resources, including light, water, nutrients, CO_2, space, pollinators, dispersers, and symbionts, varies spatially and temporally. Individual plants respond to these patterns and simultaneously modify the patterns for themselves and for their neighbors. Thus, each individual has an *effect* on, and a *response* to, its environment, including its neighbors (Jacquard 1968, Goldberg and Werner 1983). Effect and response are directly linked. Concepts such as 'effect' and 'response' and especially 'target' and 'neighbor' have been used for convenience and simplification (e.g. Goldberg 1990). In reality, individuals simultaneously affect and respond to each other. Because of differences in size, identity, and competence in resource capture and modification, they usually influence each other unequally. Therefore, *reciprocity* and *asymmetry* are common in plant–plant interactions in nature. Since both change with time, their conceptual and experimental decomposition remains a great challenge to functional plant ecology. These complex and reciprocal interactions, mediated by resources and controllers, together with certain evolved life history traits, determine the future composition of the vegetation in a given location. Therefore, a fundamental question in our approach to studying plant succession and ecosystem recovery is: *how do plants of different successional positions respond to the changing patterns of resource availability in their habitats?*

Our general research strategy combines field manipulations, controlled environments, and laboratory experiments on the individual, population, and community levels, together with field observations of patterns of community structure and field microenvironments. We believe that these approaches complement each other (Fig. 2.3) and are necessary for an indepth understanding of the behavior of plants in disturbed habitats and changing climates. As a result, our research does not fit neatly into the usual categories of physiological, population, or community ecology.

Allocation of captured resources by plants

Since the plant must operate as a balanced system, maximizing fitness requires the optimal allocation of limited resources, and allocation strategies must be under strong evolutionary control. Plant growth is frequently limited by several resources (Bloom *et al.* 1985, Chapin *et al.* 1987). However, plants are able to compensate for resource limitation by changing allocation to organs and functions most directly related to improving the uptake of the most limiting resource (Bazzaz *et al.* 1987, Rastetter and Shaver 1992).

Plants allocate biomass to belowground parts for the acquisition of water and nutrients and to aboveground parts for the acquisition of light and carbon dioxide. They also allocate resources to storage, reproduction, and defense. Differences in these allocations have been extensively studied (Bazzaz *et al.* 1987). Growth analysis techniques have facilitated this work (Evans 1972, Causton and Venus 1981, Hunt 1982, 1990). Allometric relationships avoid the confounding influence of plant size on relative allocation. Plants vary enormously in the timing and frequency of reproductive episodes and also in the number, size, and quality of the progeny (Harper 1977, Silvertown 1987). They also differ greatly in the amount of resources allocated to reproduction. The speed in which allocation patterns respond to a change in the availability of resources must be closely linked to the

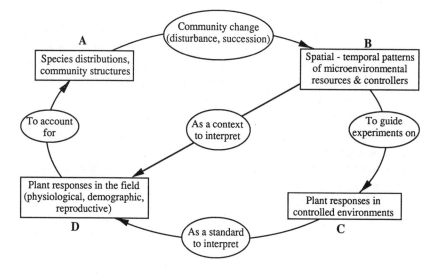

Fig. 2.3. An outline of our research strategy used in the study of plants in changing environments, combining field and controlled-environment experiments to answer questions about disturbance and succession (modified from Bazzaz and Sipe 1987).

nature of availability in the environment. It is therefore expected that the variable early-successional environment will select for species that can quickly shift their allocation. For example, increased allocation to roots in droughty situations and when nutrients are limiting is common in many early-successional plant species.

The theory of allocation was borrowed from economics by Robert McArthur and extended to plants by Harper (1967). As applied to plants, the theory assumes that an individual has a specified (and fixed) rate of supply of a resource (carbon, nitrogen, energy, etc.) that must be divided optimally between several competing structures and functions. It also assumes that what is allocated to one structure or function can not be allocated simultaneously to another structure or function. In plants, however, the assumptions of fixed resource supply and mutually exclusive allocation are not always met, because plant structures can contribute to more than one function (Fig. 2.4). In evolutionary terms, optimum allocation maximizes the relative fitness of the individual during its entire life span. Fitness is determined by total reproductive output, which is a function of the timing and frequency of reproduction (Willson 1983), and the quality of the propagules (Bazzaz and Reekie 1985). Considerable work has been done on the trade-offs between seed size and seed number (Harper

Reproductive Allocation   (RA)   **Reproductive Effort   (RE)**

Direct

(1) $\dfrac{Rr}{Tr}$ (4) $\dfrac{(Rr + Rv + Sr + Ar) - Pr}{(Tr + Sv + Av) - Pr}$

(2) $\dfrac{Rr + Rv}{Tr}$ Indirect

(5) $\dfrac{Vr - Vn}{Vn}$

(3) $\dfrac{Rr + Rv + Sr + Ar}{Tr + Sv + Av}$

(6) $\dfrac{(Vr + Sv + Av) - (Vn + Sn + An)}{(Vn + Sn + An)}$

Components of Resource Pool

Reproductive Plant   Non-reproductive Plant
Tr - Total standing pool   Vn - Vegetative size
Vr - Vegetative pool   Sn - Structural losses
Rr - Reproductive pool   An - Atmospheric losses
Rv - Vegetative biomass attributable to reproduction   (respiration/transpiration)
Sv - Structural losses from vegetative organs
Sr - Structural losses from reproductive organs
Av - Atmospheric losses from vegetative organs
Ar - Atmospheric losses from reproductive organs
Pr - Enhancement of total resource supply due to reproduction

Fig. 2.4. Components of reproductive allocation (RA) and reproductive effort (RE) and their calculations for plants (modified from Bazzaz and Ackerly 1992).

et al. 1970, Silvertown and Lovett-Doust 1993). Studies have shown that seed size varies considerably, even within populations (e.g. Thompson 1984, Wulff 1986, Sultan and Bazzaz 1993*a,b,c*), and that large seeds may confer competitive advantage (Gross 1984, Stanton 1984*a,b*, Foster 1986, Choe *et al.* 1988). In nine temperate tree species, Grime and Jeffrey (1965) showed that survivorship in the shade was positively correlated with seed size. Conversely, Shipley and Peters (1990) found a significant negative correlation between seed size and seedling relative growth rate in 204 species they examined. The mechanistic relationship between seed size, relative growth rate, and survivorship and growth in resource-limited environments requires further clarification.

Reproductive allocation (RA) is the proportion of biomass or other resources (e.g. nitrogen) found in reproductive structures at the time of final harvest. Reproductive effort (RE) is more difficult to assess than RA, since the former concerns the effort in terms of all the biomass, energy, resources, and activities that go into producing progeny. But reproductive allocation may also be inaccurate (see Bazzaz and Ackerly 1992 for a discussion). For annuals with indeterminate growth, the time of harvest may strongly influence the measurement of RA (e.g. Geber 1990). In many species in seasonal environments the termination of life occurs at different times, depending on the time of the first killing frost or the onset of a dry season. Likewise, by the time seeds are approaching maturity, the plants may have lost most of their leaves. In early-successional habitats reproduction is mostly by seed. However, in mid-successional fields clonal plants are more common and allocation to reproduction versus growth can not be easily distinguished. In long-lived iteroparous perennials there is a carryover of biomass from year to year. In late-successional trees 'standing' reproductive structures may represent a small fraction of annual production allocated to reproduction. Herbivory on vegetative parts and flowers and predation on seeds may further complicate the assessment of RA. Thus, three different measures of RA must be distinguished: (1) standing RA, the proportion of a resource contained in reproductive structures; (2) short-term RA, which is the proportional allocation of a resource over a short time integral relative to that of plant life; and (3) lifetime RA, the proportion of the total resources invested in reproductive structures over the entire life of an individual.

Reproductive effort (RE) is defined as the investment of a resource in reproduction that results in its diversion from vegetative activities (Reekie and Bazzaz 1987*c*). Several factors can decouple RA from RE. Reproduction itself may enhance resource supply, including enhanced vegetative growth, and resources can be moved between vegetative and reproductive structures. Furthermore, the reproductive structures of many species are photosynthetic. In the early-successional annual *Ambrosia trifida*, *in situ* auxiliary photosynthesis of flowers and fruits can contribute 41% and 57%, respectively, of the carbohydrate demands of male and female inflorescences. Excised pollinated female flowers can actually mature seed on their own (Bazzaz and Carlson 1979). In *Fragaria*, Jurik (1985) found *in situ* photosynthesis contributed up to 9% of carbon cost. *In situ* photosynthetic contribution is wide-ranging and does not seem to correlate with successional status. In 15 temperate forest tree species, *in situ* photosynthesis contribution ranges from 33.2 in *Acer saccharum* to 2.3 in *Quercus macrocarpa* (Bazzaz *et al.* 1979). The mid-successional *Acer rubrum* and late-successional *A. saccharum* have a similar *in situ* photosynthetic contribution of flowers and seed. Similarly, enhancement of leaf photosynthesis can be significant during reproduction. In some genotypes of *Agropyron repens*, a clonal grass of mid-successional habitats, photosynthesis during reproduction was 64% more than the rate of leaf photosynthesis prior to reproduction (Reekie and

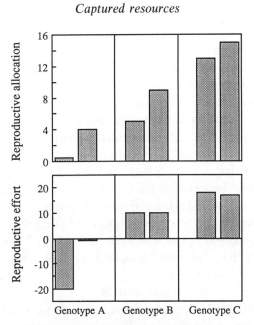

Fig. 2.5. Difference in allocation and costs among genotypes of *Agropyron repens* from fields of varying disturbance frequencies (A, B, C). Genotypes differ greatly in allocation and in costs under different resource levels. For example, genotype A has negative cost under low nutrient conditions (left) and no cost under high nutrient conditions (right) (data from Reekie and Bazzaz 1987*a*).

Bazzaz 1987*a*). Because of *in situ* photosynthesis of reproductive structures and enhancement of vegetative photosynthesis during reproduction, the cost of reproduction in terms of carbon may be negligible or even negative for some genotypes (Fig. 2.5).

It is clear then that reproductive effort is not synonymous with reproductive allocation, but is the more critical of these two measures for understanding the evolution of reproductive strategies in plants. Reproductive effort (RE) can be defined physiologically in terms of the resources invested in reproduction derived from the vegetative plant, at the same time and considering the changes in vegetative biomass resulting from reproduction (Reekie and Bazzaz 1987*a,b*). However, in order to unambiguously understand the significance of this physiological RE at a given point in the life cycle it must be related to the life-long RE of that individual. What does a certain investment in RE mean with regard to life-long reproduction? The cost of reproduction must be assessed both physiologically and demographically. What trade-offs exist between present reproduction and future reproduction? Is it always the case that copious early reproduction leads to less late reproduction? Life history theory requires that a genetic correlation exist between current and future reproduction. In *Poa annua* Law (1979) found a strong negative genetic correlation between reproductive value in successive years. Experiments with *Polygonum arenastrum* (Geber 1990) showed a negative genetic correlation between early and late reproduction but no genetic correlation between early and total reproduction. Trade-offs between current and future reproduction have been demonstrated in some species (Martínez-Ramos *et al.* 1988*b*, Snow and Whigham 1989, Primack and Hall 1990) but not in others (Horvitz and Schemske 1988). In *Agropyron repens*, for example, the physiological cost of reproduction in one season can influence survivorship and reproduction of the individual in the future. However, the cost of reproduction varies depending on genotype and resource availability (Reekie and Bazzaz 1987*a,b*). Trade-offs between reproductive activity and tree growth have been also demonstrated (Kozlowski 1971, Newell 1991).

There may be selective advantages to modifying reproductive schedules in order to minimize cost per propagule. *Plantago major*, which occurs in frequently disturbed habitats, reproduces at a small size and allocates heavily to reproduction. *Plantago rugelii*, which occurs in more stable habitats, reproduces when it is larger and allocates less to reproduction. Reekie and Bazzaz (1992) determined the effect of reproduction on growth in several genotypes of each species. In both species genotypes varied widely in the effect of reproduction on growth. Cost (reduction in growth per gram of seed capsule) increased with reproductive investment in *P. rugelii* and with plant size in *P. major*. Thus *P. rugelii* may reproduce to a lesser extent than *P. major* because cost increases with reproductive output. *Plantago major*, on the other hand, reproduces earlier than *P. rugelii* because cost increases with plant size (Fig. 2.6).

Patterns of allocation to reproduction can also vary greatly with regard to the resource being measured. Allocation of nitrogen relative to carbon varies among species from the same community (Fig. 2.7) and can be influenced by levels of other resources. In *Verbascum thaspus* RA was 40% based on biomass but varied from 5%–60% for other elements (Abrahamson and Caswell 1982).

Plants respond to a multitude of resources

Plant growth and successful reproduction depend on the acquisition, from the environment, of several required resources in balanced quantities including light, nutrients, water, CO_2, and available space (see Bloom *et al.*

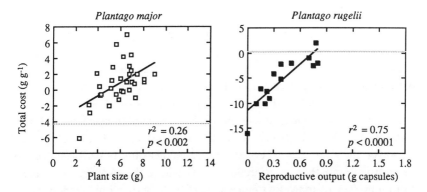

Fig. 2.6. Reproductive cost (lost vegetative growth per gram of capsules produced) in *Plantago major* from frequently disturbed and *Plantago rugelii* from less frequently disturbed habitats (data from Reekie and Bazzaz 1992).

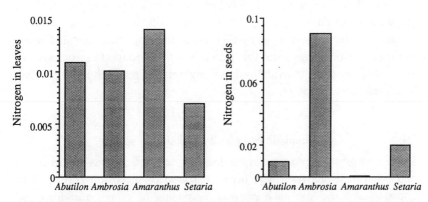

Fig. 2.7. Nitrogen percent in leaves and seeds of four early-successional annuals grown under the same soil nutrient conditions (data from Garbutt *et al.* 1990).

1985). Despite the fact that it has long been recognized that the environment is multifaceted and complex, or 'holocoenotic' (see Billings 1952), much emphasis has been placed on plant response to a single factor. In particular, plant responses and adaptations to limiting resources have been the subject of many studies in physiological plant ecology (Mooney 1991). Recently, however, there has been growing recognition that plants actually adjust the level of uptake of all critical resources so that no one is more limiting than the others (see Chapin *et al.* 1987). The commonly observed shifts in allocation to roots and shoots with changing levels of above- and belowground resources is a means of adjusting resource imbalance that can occur very quickly in fast-growing plants. Environmental complexity and the multitude of possible responses of plants to this complexity can generate great conceptual and experimental difficulties. However, under certain situations, environmental factors can be closely correlated with each other. In such cases, the response of the plants can be predicted using many fewer parameters than would be required had the resources been independent of each other. This will be discussed further in the next section.

In many physiological studies, biomass is assumed to be correlated with seed production, which is critical to explaining population dynamics. However, the relationship between these two measures of performance can be complex and there are circumstances where the two are decoupled. Minimum size for reproduction (Hartnett 1990), a threshold response that exists in many plants, is a form of decoupling the ecological implications of which are just beginning to take shape (see Bazzaz and Ackerly 1992). Resource interactions that decouple vegetative and reproductive activities are critical to the population-level response of a plant to the multiple resources of its environment. Therefore, the notions of multiple resource limitation, the response to single versus multiple resource additions, resource compensation, and the decoupling of reproduction from growth are among the most relevant and exciting areas of research in integrative, functional plant ecology.

Mechanistic complexity and simple scaling

Despite impressive progress in the last few decades, studies of succession and ecosystem recovery have been frustrated by a tension between two approaches. One approach emphasizes posing of simple questions and hypotheses. These questions have the potential of producing concrete answers that can be easily understood and quickly assimilated into the current literature. In such studies parts of systems are treated separately and even viewed independently. In contrast, the second approach considers ecological systems to be exceedingly complex, with large numbers of interacting parts, and assumes that the behavior of the whole system and its emergent properties cannot be straightforwardly understood from the study of individual parts. This view dictates that experiments be multifaceted and complex. Results of such experiments are complicated, understood only with much effort, and become integrated in the literature much later. In some instances the data from these experiments are not amenable to commonly used statistical analyses.

How to simplify this complexity and retain a reasonable approximation of reality has been a major challenge to ecology. It has been possible to use

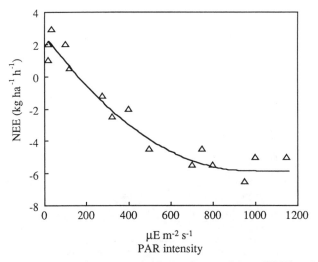

Fig. 2.8. Relationship between net ecosystem carbon exchange (NEE) and incident radiation in Harvard Forest (a temperate deciduous forest) (modified from Wofsy *et al.* 1993).

known correlations among environmental factors and among plant responses to reduce the number of parameters needed to predict some behavior of the system. For example, at the ecosystem level, there can be a strong relationship between the amount of photosynthetically active radiation (PAR) and primary productivity. Thus some measure of radiation, particularly absorbed PAR (APAR) or intercepted PAR (IPAR), would allow season-long primary productivity to be estimated (see Wofsy *et al.* 1993, Waring *et al.* 1995) (Fig. 2.8). In gaps and other early-successional habitats, mean air temperature and mean photon flux are usually positively correlated (Fig. 2.9). In temperate forests, soil respiration and CO_2 flux can be estimated from measurements of soil temperature (e.g. Peterjohn *et al.* 1993). Many new remote sensing techniques to estimate vegetational productivity and health depend on measurement of reflected light in certain regions of the spectrum. The commonly used normalized difference vegetation index (NDVI) is based on the ratio of reflected light in the wavelengths 665 and 789 nm. Leaf area index (LAI) has also been successfully used to estimate several growth parameters in forests (Waring and Schlessinger 1985). Nitrogen concentration in the foliage may be used to infer large scale ecosystem productivity (Schulze *et al.* 1994). Using such physiological parameters to infer ecosystem response is becoming more and more important in the study of global ecology, and a new field of ecosystem physiology is emerging. Currently, techniques such as controlled

Resource-response perspective

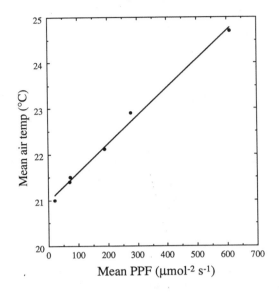

Fig. 2.9. Positive and strong correlation between light intensity and air temperature in gaps and the understory of a deciduous forest (modified from Bazzaz and Wayne 1994).

environmental chambers and open top chambers are being used in field studies. Free air carbon dioxide exposure (FACE), in which a piece of vegetation is subjected to elevated CO_2, is being tested. An alternative approach is to grow seedlings in controlled environmental facilities in ambient and elevated CO_2 environments to obtain the 'CO_2 effect.' Additionally, seedlings of the same species are grown at various levels in the canopy of natural forests to obtain the 'glasshouse effect' and measurements of canopy leaves are made *in situ* on forest trees. By knowing response among these entities, we can infer the response of canopy trees in the forest to elevated CO_2 environments of the future without subjecting the forest to elevated CO_2. Figure 2.10 illustrates these relationships. By obtaining (α) the age scaling factor (Γ) the location scaling factor, and (β) the CO_2 level scaling factor, the response of trees to elevated CO_2 can be predicted. This approach is particularly powerful if the scaling factors are close to one, such as in the case of white birch, but can be less dependable if the scaling factor greatly departs from one, such as in the case of red oak. Correlation among environmental factors can select for plant traits that are advantageous for all of these factors. For example, in high light environments, plants usually increase allocation to roots, reduce the area of individual leaves, and increase leaf thickness. These traits, which are known to be plastic in response to light alone, also cause the plants to have a high water-use efficiency and a more favorable water balance. Correlations among environmental resources can cause the improper assignment of a given plant's physiological, morphological, or behavioral response to a single resource. While these correlations can be of great value for modeling purposes, they may not contribute much to mechanistic understanding of plant response. In fact, the complexity of the plant's environment may produce counter-intuitive results with regard to the response to a single factor. Low aboveground growth in sunny environments, despite high photosynthetic rates on a leaf area basis, may be the result of much allocation to roots that forage widely for water in habitats with a severe dry season.

Early-successional plants and other colonizing species experience a variable biological environment as well. The identity, biomass, and allocation patterns of neighbors may differ, generating differing impacts on resource availability for the target plant. The intensity of herbivory and the prevalence of pathogens and symbionts can also vary greatly. By consuming, reducing growth, and killing some individuals in a neighborhood, they too can greatly modify resource availability. Because of differences in dispersal into these habitats and existence of a long-lived seedbank, early-successional plants can be recruited in a wide range of densities, from one individual to several thousand individuals per square meter. In crowded situations populations develop size hierarchies in both vegetative and reproductive performance. These and other biotic interactions influence the current resource status of a plant as well as its ability to capture new resources in the future. These size differences may have no genetic (i.e. only environmental) base (see Chapter 7). In populations of the annual *Polygonum pensylvanicum*, for example coefficients of variation in vegetative and reproductive biomass increases with population density. Density also increases the variance for fitness-related characters, such as reproductive biomass, but decreases their heritability. The same genotypes are not equally superior or inferior in all densities, and there are cross-overs in reaction norms of fitness related characters (Fig. 2.11). Simple genetic models suggest that the potential for natural selection among genotypes increases with density. However, because plants can occur in a very wide range of densities in the field, selection among genotypes can be reduced and a high level of variation maintained because there are low-density refugia for competitively inferior genotypes. Furthermore, because of generally low size thresholds for reproduction in many early-successional annuals relative to late-successional species, the effective population size (i.e. the percentage of reproducing

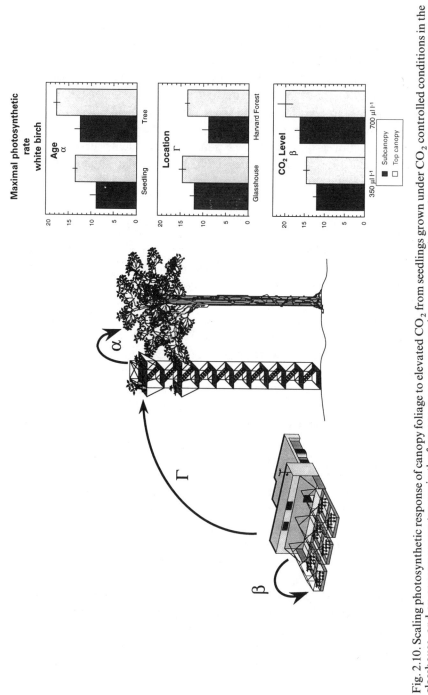

Fig. 2.10. Scaling photosynthetic response of canopy foliage to elevated CO_2 from seedlings grown under CO_2 controlled conditions in the glasshouse, and on canopy excess towers in the forest to measurements taken on canopy leaves of trees in the forest (**Bassow** and **Bazzaz** unpublished data).

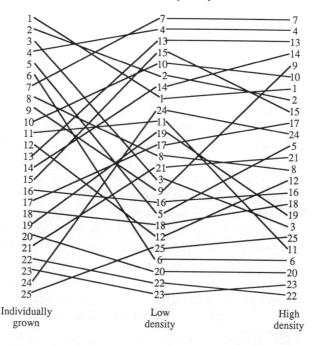

Fig. 2.11. Rank change with density of *Polygonum pensylvanicum* genotypes from a single natural population in an early-successional field (modified from Thomas and Bazzaz 1993).

individuals relative to the total number) can be larger for early-successional species, usually in dense stands. All these events can have a strong influence on the patterns of resource availability for individuals in populations.

The approach that we have taken in our research depends on the principles of the simultaneous interaction of the plant with the totality of its physical and biological environment. This interaction is mediated through resources and controllers. We consider the plant to be responding to complex and changing levels of resource fluxes in a coordinated way and adjusting its activities and functions such that it grows and reproduces in its habitats. These habitat factors have led to the evolution of life history strategies that either optimize relative life-long fecundity in the site, or permit a rapid life cycle and subsequent escape to new sites where the environment is suitable for growth and reproduction. The key concepts that we consider in this complexity are multiple resources and controllers, resource capture, congruence, capacitance, and reciprocal influences in a spatially and temporally changing environment.

Resources and responses: a phytocentric view of community dynamics

Experimentally, succession and ecosystem recovery are often studied by examining plant variables such as weight, dominance, diversity, or seed number – the final higher level outcomes of many other phenomena. In changing habitats, interactions occur among individuals that themselves greatly vary in their life history strategies. For example, late-successional plants must compete successfully with both juveniles and adults of early-successional plants, and with themselves both as juveniles and as mature individuals. Early- and late-successional trees will compete with each other, and as seedlings, saplings, and sprouts will compete with the herbaceous plants of early-successional fields and pastures. Thus in successional habitats the stage is set for interactions among individuals that can differ in identity, size, and stage of maturity. While there can be 'winners' and 'losers' among interacting individuals, in succession there are no consistently winning and losing species, so long as each grows, reproduces, and disperses its propagules when it matures, whether that takes one season for an early-successional annual or 50 years for a late-successional tree.

Currently, disturbance is viewed as a mechanism for initiating succession through biomass destruction (e.g. Grime 1979), increased space availability (Sousa 1984), and change in resource availability patterns (Bazzaz 1983). Plants acquire and allocate resources through whole-plant integration of physiology, modular functional morphology, and architecture. They interact with each other through the modification of aboveground and belowground fluxes in physical and chemical factors of the environment. We believe that succession is most profitably studied through differential recruitment, resource acquisition and allocation, growth, and mortality of plants that are using shared environmental resources needed to grow and reproduce (see Bazzaz and Sipe 1987). Therefore, the approach we have taken in our studies on succession and ecosystem recovery is based on patterns of resource availability as affected by disturbance events, and the way species respond as individuals and in populations to resource patterns: a resource-response perspective. Change can be most clearly understood by viewing the interactions among plants through their influences on the physical and chemical fluxes in their common environment. The development of alternative resource-use strategies by plants through their evolution is assumed to be driven largely by the relative costs and benefits of alternative strategies as they ultimately affect differential fitness. Plant–animal interactions, such as pollination, dispersal, seed predation, and herbivory, which are also important to ecosystem recovery, can be viewed in terms of cost-benefit accounting of limited resources (e.g. Mooney and Gulmon 1982, Bazzaz *et al.* 1987). Furthermore, since successional change involves activities at the individual, population, community and ecosystem levels, its study can be unified through emphasis on the resource fluxes in various microsites and the physiological ecology of resource capture and use by interacting individuals. Together with dispersal patterns and the unfolding of life histories, these interactions determine successional trajectories.

3

Community composition and trends of dominance and diversity in successional ecosystems

Ecologists have hotly debated whether succession is directional, leading to predictable stable communities, or a chance process without clear patterns and definite endpoints (see McIntosh 1980, Pickett *et al.* 1989). With the current and near universal understanding that most recovering ecosystems are patchy, it has become apparent that the answer to this question depends on the scale considered. On a 'patch' scale, chance events such as dispersal patterns and their coincidence with environmental factors condusive for recruitment are critically important in determining the composition of the vegetation in the patch. Various patches can differ greatly in their composition. They may have different occupants initially recruited from among available species that can function in an open environment. However, on a larger landscape or regional scale, as these patches average out over space and time, the general physiognomy of the vegetation may appear uniform. Despite the importance of the initial conditions, including the identity of occupants, patches of dissimilar vegetation within a region can eventually converge in species composition, if allowed enough time without disturbance. The Clementsian views of succession, which have been severely criticized by many authors, are based on a regional rather than on a patch scale and do not clearly recognize the prevalence of repeated disturbances. Also, trends in ecosystem recovery such as biomass accumulation, nutrient circulation, and species diversity are usually considered at a larger than a patch scale (Odum 1969) but smaller than a regional scale. Many of these differences in views would become less strong if scale were unambiguously identified and considered.

Patterns of dominance

Many studies in the eastern deciduous forests of the United States (e g. Oosting 1942, Bard 1952, Quarterman 1957, Bazzaz 1968, 1975, Horn

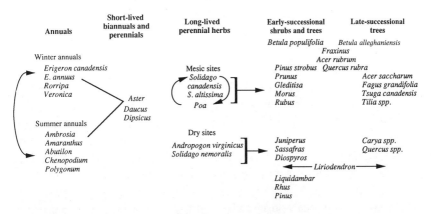

Fig. 3.1. Idealized flow diagram of species and life forms in temperate successions in the eastern United States.

1971, Pickett 1980), and in other temperate forests in Japan (e.g. Hayashi 1977, Numata 1990) and Central Europe (e.g. Bornkamm 1984, 1985, 1988, Schmid and Stöklin 1991) have established that the general trends of species dominance are similar in these regions. This is particularly true on abandoned agricultural lands and where patchiness is not intense. In fact, these recovering ecosystems, despite their different geographic locations, share several cosmopolitan herbaceous species that play comparable ecological roles. Prominent among these species are the winter annual *Erigeron canadensis*, the summer annual *Ambrosia artemisiifolia*, and the clonal mid-successional species *Solidago canadensis*. These species all possess physiological and demographic features that make them well suited to dominate these habitats, as will be shown later in the book.

In many temperate forests of eastern North America, a generalized pattern of dominance during succession from abandoned fields can be described in five phases (Fig. 3.1).

1. During the first year after soil disturbance, annual plants dominate. Many of these species are weeds commonly found in agricultural fields and on roadsides, and their seeds are usually present in large numbers in the soil prior to the initiation of disturbance. Among these plants two families, the *Cruciferae* and the *Compositae*, are especially important. The annuals of early-successional fields can be of two groups: the winter annuals and the summer annuals. The winter annuals are usually small-seeded, germinate in late summer and fall, spend the winter as

rosettes near the ground, and bolt, flower, and set seed in the following summer. The most widespread species in this group are *Erigeron canadensis* and *Erigeron annuus*, which can form near pure stands. In contrast, the summer annuals usually have relatively large seeds that persist in the seed bank, and most of which require overwintering (cold treatment) for germination. Their seeds disperse in late summer and early fall. They generally germinate in the spring and dominate the fields in the summer, especially in the absence of the first group. *Ambrosia artemisiifolia* is the most representative and widespread of this group, but other species, such as *Chenopodium album*, *Amaranthus retrofluxes*, *Polygonum pensylvanicum*, and *Abutilon theophrasti*, are also prevalent. In North America and elsewhere, this group is usually a mixture of native and introduced species. On extremely poor soils, *Digitaria sanguinales* is important, and in some highly eroded locations, *Diodia teres* is dominant. Several annual grass species, especially *Setaria faberi*, are also present.

2. The second year of succession is usually dominated by short-lived perennial herbs that may actually recruit during the first year, but become prominent only in the second year because of their relatively slow growth rate. The leading example of this group is *Aster pilosus*. Some winter annuals, such as *Erigeron annuus*, can also be present in these fields. However, many of the summer annuals are eliminated (e.g. *Abutilon*) or are present as suppressed seedlings (e.g. *Ambrosia*), which suffer great mortality in competition with the established perennials. These annuals contribute very little to the biomass of the community, but can still contribute appreciably to the seed bank because of their low size threshold for reproduction and relatively high reproductive allocation.

3. The third phase of recovery is usually dominated by clonal herbs that can persist for several years. The demise of this phase is heterogeneous and is caused by the patchy invasion of early-successional shrubs and trees that act both as foci for seed dispersal (Fig. 3.2) and as modifiers of the microenvironment beneath them, which may in turn control future invasion. This third phase is dominated by members of the genus *Solidago*, with *S. nemoralis* on nutrient-poor soil, *S. altissima* in the midwest, and *S. canadensis* and (to a lesser extent) *S. gigantea* in the east on moderately poor soils. *Solidago altissima* and *S. canadensis* are becoming very important in abandoned fields in Central Europe (Zwölfer 1976, Schmid and Stöklin 1991). *Solidago altissima* is also becoming widespread on disturbed ground in Honshu Island in Japan, where it grows unusually tall. Grasses may also be common in this phase of

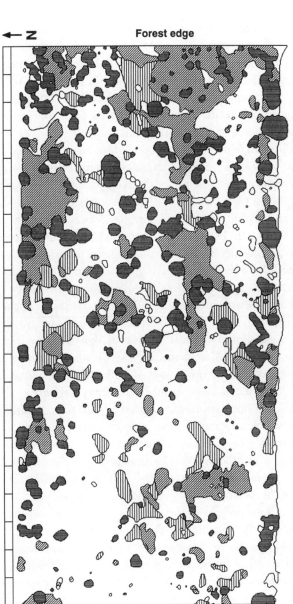

Forest edge

N ←

Fig. 3.2. Spatial distribution of individuals and patches of trees in a mid-successional field with an herb layer (in white) made up of *Solidago* and *Poa* and trees of *Fraxinus*, *Prunus*, *Gleditsia*, and *Morus*. Proximity to seed sources (a mature phase deciduous forest) is evidenced by higher density of the trees on the right-hand side (data of Burton and Bazzaz unpublished).

succession, and, like *Solidago*, are patchily distributed. Among the grasses in the southeastern United States, *Andropogon virginicus* is very common, especially on poor soils deficient in phosphorus. In the midwest, *Poa pratensis* dominates, as *Agropyron repens* does in the north, especially on sandy soils. *Aristida dichotoma* prevails on highly eroded soils in oldfields in many locations in the eastern United States. In many successions, for reasons which are still unknown , the grasses and *Solidago* can replace each other cyclically. Still, neither persists in deep shade, and both are ultimately replaced by sun-adapted early-successional shrubs and trees in the fourth stage.

4. Shrubs common in mid-successional fields usually include species of the genera *Rhus* and *Rubus*. Early-successional trees include the clonally spreading *Sassafras* and *Diospyros*, the evergreen *Juniperus* and *Pinus*, and the nonclonal trees *Ulmus alata*, *Gleditsia triacanthos*, *Liquidambar styraciflua*, and several species of *Crataegus*. *Prunus serotina* is common in the west and north. Depending on geography, the early-successional tree invaders of mid-successional fields can also include *Betula populifolia*, *Populus grandidentata*, *Prunus pensylvanicum*, *Acer rubrum*, and *Nyssa sylvatica*. Tree invasion is highly conditional on proximity of seed sources and on dispersal patterns. Except for *Pinus taeda* and *Liquidambar*, which differ in their soil moisture tolerance (see Tolley and Strain 1984), there are no clear physiological differences that strongly dictate the sequence of arrival of different species. Clearly, all of these species can tolerate some degree of limitation of soil moisture and nutrients, conditions that are characteristic of open, sunny habitats.

5. Late-successional habitats are dominated by various combinations of tree species, which may include one or more species of *Acer*, *Aesculus*, *Betula*, *Carpinus*, *Carya*, *Fagus*, *Liriodendron*, *Populus*, *Quercus*, *Tilia*, *Tsuga*, *Ulmus*, and others. *Assimina triloba*, *Cercis canadensis*, *Cornus florida*, and *Lindera benzoin* may be found in the understory. Small shrubs, such as several species in the family *Ericaceae* and genus *Viburnum*, are also present in the understory, especially in the east. A species-rich, mostly perennial, herbaceous ground flora with showy flowers predominates on nutrient-rich, moist sites, and ferns usually predominate on relatively nutrient-poor sites. Some tree species, such as *Fraxinus americana* and *Liriodendron tulipifera*, may appear early in tree succession but persist in some mature-phase forests. *Liriodendron* in particular is shade-intolerant with a fast growth rate and a long life, and can persist on a site for several hundred years (Shugart 1984). Because of the great diversity of late-successional species (especially within the

genus *Quercus*), many different combinations are possible within a region and in different locations several community types can be identified. The feature that ties these disparate communities together is the predominance of deciduous trees. Because of the long life of these trees, there are no field observations long enough to determine precisely their successional replacement patterns under different circumstances.

Despite much research on the life history of many forest trees (see Kozlowski *et al.* 1991), patterns of tree-by-tree replacement in various successions remain unclear, especially during the later phases of succession (see Paulson and Platt 1989). In fact, in late succession, a species' successional status is more relative than absolute. Thus, especially for trees, it is more appropriate to designate species' successional status as earlier or later in relation to other species. Overly rigid attempts to classify species as mid- or late-successional have generated confusion in the literature. In the forests of New England, USA, *Betula alleghaniensis* (yellow birch) has been called an early-, mid-, and late-successional tree by various authors. Similarly, *Quercus rubra* (red oak) has been called early- and late-successional in these forests. Experience has shown that relative to white and black birches, yellow birch is late-successional, but relative to sugar maple and especially to beech and hemlock, it is early-successional. Furthermore, the relative position of species on a successional gradient may vary among different regions.

Although late-successional tree species can be classified by their shade tolerance (see Burns and Honkala 1990) and the depth of the shade they cast (Horn 1971, 1976), they replace each other less predictably than do species of early- and mid-successional habitats. Apparently, dispersal patterns and stochastic events can override differences in physiological response to environmental resources. Thus, these replacement events are governed by stochastic phenomena and possibly by subtle yet important differences in microsite characteristics that favor certain species over others. The subtle differences among patches may themselves arise from long occupancy by certain species, which leave their own 'legacy' on the site in the form of specific root exudates, litter quality and chemistry, mycorrhizal and microbial associations, pathogens, and seed predators. These site modifications may encourage the same species to reoccupy the site (Fig. 3.3) or may inhibit recruitment of the same species again, as species-specific seed predators and pathogens concentrate on the site (Janzen 1970, Connell 1971). The balance between 'site familiarity' and 'inhospitability' may change from location to location and year to year. Such species

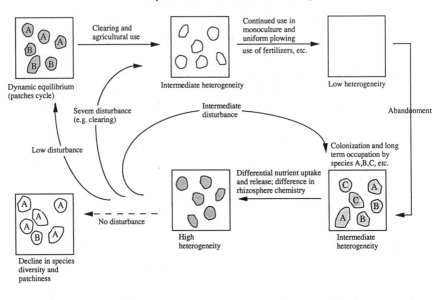

Fig. 3.3. Graphical representation of the relationship between disturbance severity, site heterogeneity, colonization, and species legacies (modified from Bazzaz 1983).

legacies can play an important role in community structure, but their nature and the direction of their influence remain largely unexplored.

Patterns of diversity

Despite great interest in biological diversity because of its presumed relationship to the health of ecosystems and to community organization (see Huston 1994), it is still unclear how to assess diversity and how it functionally contributes to community properties. Usually species diversity is measured taxonomically, but non-taxonomic descriptions of plants are used by some European ecologists (e.g. Barkman 1988). An index of diversity based on parameters other than taxonomic diversity has the potential to give more insight into the processes of regeneration and coexistence (see Grubb 1977). Diurnal patterns of nectar availability and flower color, for example, may be more important to the diversity of the pollinator community than the taxonomic diversity of the plant species (Fig. 3.4). Flower color may also provide the best assessment of diversity from an aesthetic point of view (Rathcke and Lacey 1985). Despite the difficulties in identifying clearly what entities should be used to assess functional biodiversity in ecosystems (see Wayne and Bazzaz 1991), the

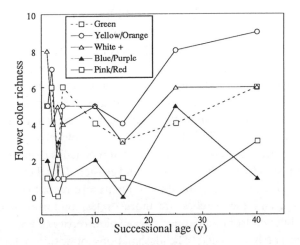

Fig. 3.4. Flower color diversity in midwestern oldfield succession. Trends are very different from those of plant species diversity (modified from Wayne and Bazzaz 1991).

subject has been very important in community studies and has been extensively considered by several authors (see Whittaker 1972, May 1976, 1978, Huston 1994), and trends in species diversity for a variety of successional ecosystems have been proposed.

Customarily, the diversity of plant species in an ecosystem has been used as a surrogate measure of total ecosystem diversity. It is assumed that plant species diversity is positively correlated with structural as well as chemical diversity, which are assumed to be positively correlated with animal diversity (MacArthur 1958, Whittaker 1975). Loucks (1970) proposed that early-successional habitats have relatively low species diversity because only a limited subset of species are adapted to the 'harsh' environment of these habitats. Stable, late-successional habitats are also relatively species-poor because competition among species in these systems leads to the elimination of competitively weak species. In contrast, mid-successional habitats should exhibit maximum species diversity, as early-successional sun-adapted species are found together with late-successional shade-adapted species. From a different perspective, the intermediate disturbance hypothesis (Connell 1978, Sousa 1979, 1984, Denslow 1980) reaches parallel conclusions. With frequent disturbance, only a few tolerant species are present, and with very infrequent disturbance, the forces of competitive exclusion eliminate many of the species. With intermediate frequency of disturbance, the number of species present in the community is at its maximum.

This relationship between species diversity and succession has been observed in many regions (e.g. Bazzaz 1975, Tramer 1975, G. M. Woodwell in Whittaker 1975, Mellinger and McNaughton 1975, Tilman 1988). Like most community attributes, species diversity and its trends with succession are strongly scale-dependent. Contrary to general theoretical expectations about species/area relationships, species number in successional habitats rarely rises smoothly and asymptotically with an increase in area. Patchiness in the distribution of resources and species can generate abrupt changes in species/area relations. This is especially noticeable in mid-successional habitats that are often characterized by heterogeneous dispersal patterns, clonal expansion from foci of invasion, and erosional patches with different resource levels from the surrounding soil matrix (Bazzaz 1975). If a field is very patchy, and the patches are more or less uniform internally but distinctly different from each other, overall biodiversity can be high despite low within-patch diversity. The designations of within-patch diversity (alpha), between-patch diversity (beta), and overall diversity (gamma) of Whittaker (1972) recognizes these scales. Species/area curves can be informative in delineating patchiness. As area increases, there is usually an increase in species number until an asymptote is approached. An abrupt increase in species number with further increase in area indicates a change in patch or community type. Species/area curves can therefore be used as a first approximation to identify different patches or community types and transitions from one patch to another in successional ecosystems.

Dominance–diversity relationships

Theory suggests that the structure of dominance among species in the community also changes with succession (Whittaker 1972, May 1976, 1978). In early-successional habitats with only a few species present, a geometric distribution in the dominance of species is expected. The most dominant species contributes much of the total community biomass and/or numbers to the community and uses a large fraction of the available resources. The next most dominant species uses a large fraction of the remaining resources, and so on down the line of dominance. In contrast, late-successional habitats exhibit a lognormal distribution of dominance with a high equitability in resource use among a few species. These sorts of distributions and patterns of dominance have been found in many successional sequences (Fig. 3.5). Within a region, however, there can be large differences in dominance–diversity relationships, depending on soil type and time of initial disturbance of the soil (Fig. 3.6). In the Illinois fields, patterns of diversity during succession indicate that communities that

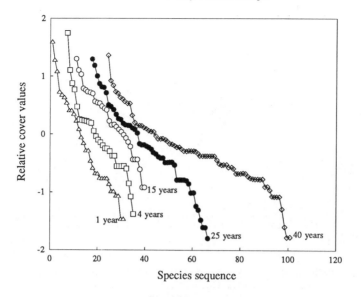

Fig. 3.5. Patterns of dominance diversity during succession in fields of up to 40 years of succession (modified from Bazzaz 1975).

initially developed on less fertile soil are generally more diverse than communities that developed on the inherently rich prairie soil, despite similarity in total plant biomass (Fig. 3.7). Furthermore, species diversity on a patch scale is influenced by soil fertility in accordance with the 'Suchatchev Effect'. That is, fast growth when soil resources are plentiful should lead to early initiation of competitive interactions, accelerated mortality of a large number of individuals and species, and a reduction in species diversity.

In some successions, certain stages may be strongly dominated by one species whose life history attributes allow it to perform well. In these cases, species diversity may be greatly reduced irrespective of the phase of ecosystem recovery. In the eastern United States, the grass *Andropogon virginicus* strongly dominates many fields after 4 to 10 years of succession, causing a decline in species diversity during that period (Bazzaz 1975). Similarly, in a 20-year study in the Illinois fields, diversity generally increased with time, with a dip during the period of strong dominance by *Solidago* (Fig. 3.5). The occasional presence of an unusually strong dominant, such as the giant ragweed *Ambrosia trifida* in oldfields, can lower species diversity by an order of magnitude (Abul-Fatih and Bazzaz 1979*a*). In contrast, in fields on very nutrient-poor, sandy soil in Minnesota, there

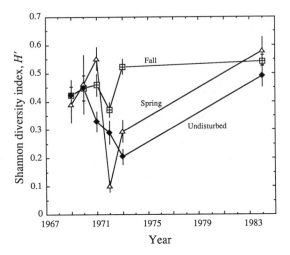

Fig. 3.6. Patterns of diversity in succession on adjacent experimental plots differing in seasonal time of disturbance. Declines are associated with the strong dominance of *Solidago*.

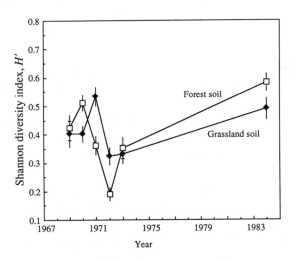

Fig. 3.7. Relation between plant species diversity and soil type in adjacent plots showing strong decline in diversity during the dominance of *Solidago*.

was a positive relationship between species richness and field age (Inouye *et al.* 1987). In most successions, biomass accumulation increases with succession (Odum 1969, Bormann & Likens 1979). For example, in the Hubbard Brook ecosystem in the northeastern United States, despite the

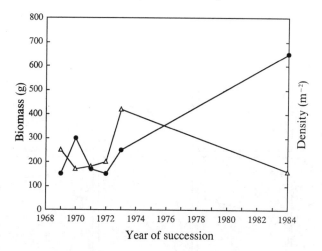

Fig. 3.8. Changes in plant biomass and density during 16 years of succession on experimental plots in the midwestern United States.

initial use of herbicides to suppress regeneration, primary productivity increased exponentially, then linearly, and by year 20 it attained 38% of mature phase productivity (Reiners 1992). Furthermore, aboveground nutrient pools accumulated faster than biomass with potassium, phosphorus, magnesium ions, and nitrogen reaching 32%, 44%, 42%, and 29% of mature phase forest, respectively (Reiners 1992).

Biomass versus density

Despite the fact that community ecologists have long recognized the importance of both number of individuals and the spatial distribution of these individuals, research in successional communities has emphasized biomass as the measure of change, possibly because it is easier to estimate. There are, however, many cases where biomass and density of a species can have very different patterns (Fig. 3.8). For example, individuals of *Ambrosia* recruited in the first year produce copious seeds that can germinate in large numbers in the second year. However, because of the strong performance of the winter annuals, and their great advantage in gaining carbon during winter, the large number of recruited individuals of *Ambrosia* remains small and their biomass is quite low. Based on an analysis of over six years of patterns of biomass and density in relation to soil type and time of soil disturbance in 99 species of higher plants in the experimental plots we found:

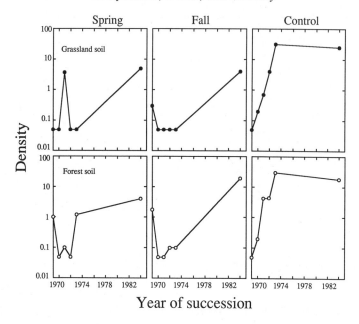

Year of succession

Fig. 3.9. Change in *Solidago* density over 12 years of succession in relation to soil type and time of soil disturbance.

1. Several annuals, notably *Ambrosia artemisiifolia*, *Setaria faberi*, and *Chenopodium album*, had peak biomass in the first year but peak density in subsequent years.
2. In contrast, *Aster pilosus* reached peak biomass and peak density in the same year (year 4). *Aster* density responded only to the time of initial soil disturbance, whereas its biomass and density responded to both time of soil disturbance and soil type.
3. With regard to biomass, 40% of the species responded to successional time, 18% to the time of initial soil disturbance, and 9% to soil type. For density, 50% responded to successional time, 24% to initial soil disturbance, and 10% to soil type.

Despite these differences, however, density and biomass of most plant species in this system appear to be more influenced by successional time than by initial soil type or seasonal timing of disturbance (Fig. 3.9). These results support the notion that initial conditions, while important to general trends of succession, are less important later in succession and do not seem to greatly modify successional trajectories. This is in agreement with the Clementsian view of convergence in regional successions. Critical tests of the role of initial conditions will require long-term observations and large-scale experiments.

Life forms in succession

Dominance of various life forms also changes with succession. In terms of biomass, annuals and biennials decline quickly, and within a few years perennials predominate. Forbs usually decline in the face of competition with graminoids and later with woody plants. Trends in dominance of native versus exotic species are not clear and are likely to differ in various successional trajectories. However, it is expected that unless disturbance is extensive and severe, and most native plants have been replaced by European species, exotics will be less important in late-successional ecosystems (Fig. 3.10). Relative to early-successional habitats, late-successional ecosystems are expected to have low invadability because of their higher species packing (see Bazzaz 1986, Vitousek 1986). In such cases there is less unoccupied resource space for exotics to establish themselves.

Clonal plants dominate mid-successional habitats

Clonality is an important plant strategy found in many vegetational types, and may play an important role in recovering ecosystems. Clonal plants are common in mid-successional habitats but are usually restricted to the understory of mature-phase temperate forests. In most secondary successions, the herbaceous clonal composites, especially members of the genera *Aster* and *Solidago*, dominate in mid-successional fields until these species are replaced by shrubs and trees. Members of the clonal shrub genera *Rubus* and *Rhus* are also common in mid-successional communities. Some of the widespread early-invading trees in oldfield successions (e.g. *Sassafras* and *Diospyros*) are also clonal. In many fields, these species form monoclonal patches, which create shady microenvironments underneath them that are hospitable to some late-successional tree seedlings. Therefore, the patterns of clonal distribution in the field can greatly influence the spatial distribution of late-successional trees.

Because their propagules can easily disperse into disturbed sites, clonal plants can also be present in fields during early succession. *Sassafras* and *Diospyros*, in fact, may already be present in the fields as fragments before the fields are abandoned. The dual strategies in some species of clonal spread (vegetative propagation) and sexual reproduction by seed ensure

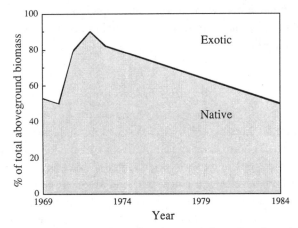

Fig. 3.10. Change in the proportions of native versus introduced species in oldfield succession in the midwestern United States.

that these species can both colonize newly opened habitats and enlarge their areas of influence once they are recruited. The interdependency among ramets of a clone is a critical component in their spread in a given location. Clonal integration facilitates the movement of photosynthates, nitrogen and other nutrients, water, and hormones across the entire genet (see Hutchings 1988, Schmid 1992), even in a patchy environment.

Despite similarities in growth habit, clonal plants differ from each other in some characteristics important to their role in ecosystem recovery. Clonal plants can differ in the length of their rhizomes, the length of time their ramets remain attached to each other, the degree of integration among ramets of the same individual genet, their growth response to different competing neighbors in terms of genet architecture, biomass allocation and biomass production, and response to herbivory. In many situations, interconnections among sister ramets of a genet (clone) can involve both cooperation and competition for resources. To study the conditions that foster cooperation or competition, we used three species of *Solidago* (*S. altissima, S. canadensis, S. gigantea*) and three species of *Aster* (*A. lanceolatus, A. longifolius, A. novi-belgii*). As mentioned earlier, *Solidago* usually replaces *Aster* in temperate deciduous forest successions.

Although the three species of *Solidago* are closely related taxonomically, they differ from each other in several ecologically important traits. Shoots of *S. canadensis* have many small leaves with high photosynthetic rates, grow quickly, flower early, and invest most of their resources in seed production. Shoots of *S. altissima* and *S. gigantea* usually have a larger leaf

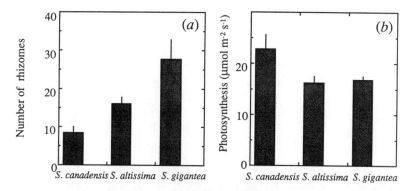

Fig. 3.11. (a) Differences in rhizome production among *Solidago canadensis*, *Solidago altissima*, and *Solidago gigantea*. (b) Photosynthesis in *S. canadensis*, *S. altissima*, and *S. gigantea* grown under the same field conditions.

area, and fewer but larger leaves with lower photosynthetic rates. These species also grow more slowly, delay flowering, and invest more resources in the production of rhizomes (Fig. 3.11). Perhaps the most important difference among the three *Solidago* species is the length of their rhizomes. *Solidago canadensis* has short rhizomes (5 cm) and forms compact monoclonal patches with high shoot (ramet) density within genets. *Solidago altissima* and *S. gigantea* have longer rhizomes (up to 20 cm) and lower within-genet shoot densities. The three species of *Aster* are morphologically similar; they all have long rhizomes (up to 100 cm or more). Therefore, in the field different *Aster* genets intermingle greatly, forming uniform stands of high density among genet clones but low density within genet clones.

Solidago altissima, whose rhizome connections remain intact for up to four seasons, can persist in successional fields for 75 years (Werner *et al.* 1980). To investigate the degree of physiological interdependence among ramets, rhizome connections of selected genets were severed in the field at different times during the growing season. Ramets severed from their parental clones experienced reduced survivorship, growth, and flowering activities relative to the controls. The earlier the severing occurred, the more severe were the effects on the severed ramets (Fig. 3.12). Thus, newly developed ramets were initially dependent upon their parental clone during emergence and establishment, but became progressively less dependent as the growing season progressed. When one of the ramets within a connected pair was shaded to reduce light levels to only 10% of full sunlight, the rate of photosynthesis in the unshaded member significantly increased (Fig. 3.13) and the survivorship of the shaded, connected ramet was prolonged relative

Composition, trends, and diversity

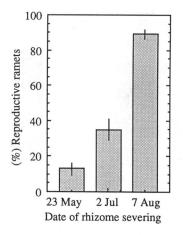

Fig. 3.12. Reduction in the number of reproductive ramets with time of severing of rhizomes from their parents in *Solidago altissima* (modified from Hartnett and Bazzaz 1983).

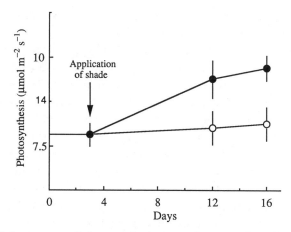

Fig. 3.13. Enhancement of photosynthetic rates when one of a pair of connected ramets is experimentally shaded (compensatory photosynthesis) (modified from Hartnett and Bazzaz 1983).

to the shaded, severed ramets. Thus, disadvantaged ramets in a connected clone receive photosynthate (and perhaps other materials) from their advantaged sisters (Hartnett and Bazzaz 1985a). Severing rhizome connections produced different results in *Solidago gigantea* and especially in the three species of *Aster*, which have longer rhizomes than *S. altissima* (Schmid and Bazzaz 1987). The growth in terms of biomass, height, stem diameter, and number of modules was only slightly influenced by severing

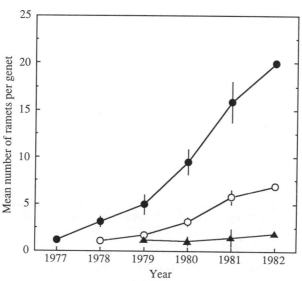

Fig. 3.14. Dominance of early-recruited and failure of late-recruited genets of *Solidago altissima* expressed as mean number of ramets produced per genet, over a five year period in the field (modified from Hartnett and Bazzaz 1985*a*).

rhizome connections of the less-integrated *Aster* species. Furthermore, *Aster* genets, which in the field usually intermingle with neighbors, performed better in mixtures than in pure stands, while the opposite was true for *Solidago canadensis*. There was also a relationship between allocation patterns and the length of rhizome (degree of integration). With increased density, *Solidago* allocated more biomass to the belowground perennating rhizomes than to the shorter-lived shoots. In contrast, *Aster* allocated more to shoots than to belowground rhizomes. These results suggest that integration is more important to clonal plants with compact genet architecture (short rhizomes), like *Solidago*, and less important to clonal plants with spreading genet architecture (long rhizomes), like *Aster*.

Recruitment of *Solidago* and other small-seeded clonal plants in the oldfields is largely the result of a single episode. In successional fields, genet recruitment begins in the third year after abandonment, and despite the continuing enormous seed input, recruitment in the following years is very small (Fig. 3.14). Early-arriving genets grow much faster than late-arriving genets and produce many more ramets. Genet mortality is also much higher in the late-arriving than in the early-arriving recruits (Fig. 3.15). As a result, genet density becomes somewhat fixed by the fourth year and remains stable for several years. In contrast, because of the continued

Composition, trends, and diversity

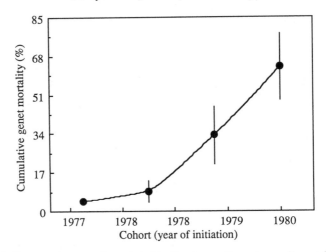

Fig. 3.15. Increase in mortality related to the times of genet recruitment in a field population of *Solidago altissima* in a mid-successional field (modified from Hartnett and Bazzaz 1985*a*).

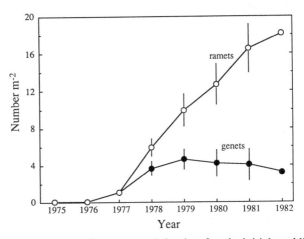

Fig. 3.16. Increase in ramet (but not genet) density after the initial establishment of *Solidago altissima* in the field (modified from Hartnett and Bazzaz 1985*a*).

growth of the early-recruited genets, ramet density increases for several years after the initial establishment (Fig. 3.16). Thus, the genetic diversity of *Solidago* in a given field is determined by the identity of the early recruits. The interdependence among ramets within clones and their ability to integrate environmental heterogeneity may buffer against patch-specific localized selection and help maintain the initial genetic diversity of *Solidago*.

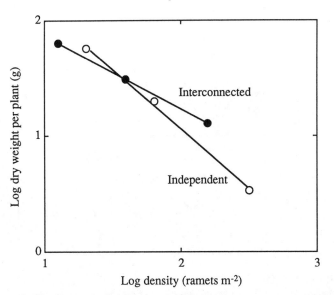

Fig. 3.17. Weight density relation for connected and disconnected ramets of *Solidago altissima* (modified from Hartnett and Bazzaz 1985*c*).

Solidago and other clonal plants have multiple levels of population regulation. In response to density stress, these populations can restrict birth and increase death at the level of the genets, ramets, or leaves. Experiments show that in independent ramets of *Solidago altissima*, increasing density results in a reduction in leaf population growth rates, ramet biomass, the number of flower heads, and reproductive allocation. Interconnected ramets of older genets show similar but smaller responses to density, suggesting that interconnected ramets are less sensitive to density stress. Ramet weight in relation to density does not conform to the 'law of constant yield' predicted for populations of individual plants (Fig. 3.17). This is further evidence that *Solidago* ramets are dependent on each other (Hartnett and Bazzaz 1985*b*). Because of their generally short rhizomes, *Solidago altissima*, and especially *Solidago canadensis*, form compact genets of connected sister ramets and expand somewhat uniformly in all directions to create pure, essentially circular clones several meters in diameter, effectively resisting invasion by other species.

Expansion of clonal plants into adjacent vegetation in mid-successional habitats is influenced by the degree of integration among sister ramets in a genet and the length of time they remain connected. Connected ramets, because they share resources, are able to integrate various patch types and

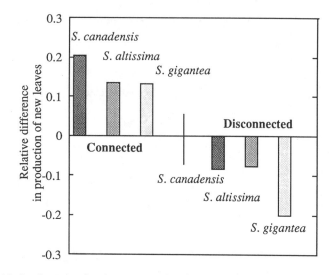

Fig. 3.18. Leaf production in connected pairs and leaf death after defoliation in disconnected pairs of *Solidago canadensis*, *Solidago altissima* and *Solidago gigantea* with increasing rhizome length and decreasing degree of integration (modified from Schmid *et al.* 1988).

therefore can expand nearly equally into all adjacent patches irrespective of the identity of neighbors. Hartnett and Bazzaz (1985*b*) have shown that in *Solidago altissima* in mid-successional fields, disconnected ramets grew differentially in different patches, but connected ramets grew equally well in patches with different neighbors. The expansion of clonal patches into adjacent vegetation also depends on the length of their rhizomes. Compact genets with short rhizomes are expected to advance in a single front, whereas genets with long rhizomes expand opportunistically and unequally into adjacent patches (Schmid and Bazzaz 1992).

The degree of integration among connected ramets of a genet can also influence their response to herbivory. Highly integrated genets with short rhizomes should suffer less from defoliation than do less integrated or completely disconnected ramets. We tested this hypothesis on the three *Solidago* species differing in their degree of rhizome integration. The connections between sister ramets were severed in half of the clones and left intact in the other half, and 50% of the leaf area of selected ramets was removed to simulate herbivory (Schmid *et al.* 1988). Isolated shoots suffered most from defoliation, connected shoots suffered least, and the response was exacerbated in high density (Fig. 3.18). The results suggest that partly defoliated shoots receive support from their non-defoliated connected sister ramets. When most of the root system is removed, shoots

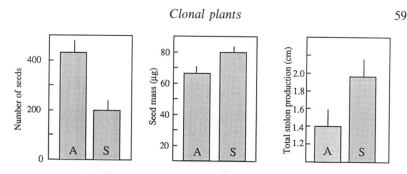

Fig. 3.19. Differences among asexual (A) and sexual (S) individuals of *Antennaria* as seed number, seed mass, and stolon production (modified from Michaels and Bazzaz 1989).

quickly restore the equilibrium between aboveground and belowground parts by redirecting resources. Thus, parts of the rhizome that are shared by actively growing shoots act as a source of material for compensatory growth and therefore as a buffer against herbivory (Schmid *et al.* 1990).

The clonal herb *Antennaria parlinii* (family Asteraceae) is stoloniferous and can reproduce by seed either sexually or apomictically. Sexual populations are dioecious, and the male and female flowers are easily distinguishable. Apomictic individuals are also easily distinguishable from sexuals. The life cycle is as follows: seeds germinate in the fall, rosettes grow and overwinter, ramets are produced in the spring, new rosettes develop at the tip of stolons, and seed dispersal by wind occurs in late spring and early summer. While sexual and apomictic populations can be found intermixed with each other, apomictics are more common in oldfields, while sexuals are found in somewhat later-succesional open woodlands (Michaels and Bazzaz 1986). In the field, apomictic plants produce more seeds per inflorescence, but the seeds are smaller than those produced by sexual plants (Fig. 3.19). Seedling survivorship is higher in sexual than in apomictic individuals. Apomictic plants will colonize new sites because of their greater potential for dispersal. Under controlled conditions, reproductive biomass of apomictic plants is consistently greater than that of sexual populations. The percentage allocation to reproduction increases much more in apomictic plants when resources (light and nutrients) are added (Fig. 3.20).

On a within-population level, the response to these resources was relatively homogeneous accross the gradient for apomictic plants, whereas the response of sexuals was more variable (Michaels and Bazzaz 1989). Although both groups did not differ greatly in response breadth on these gradients, the within-genotype response was three times as much in sexuals

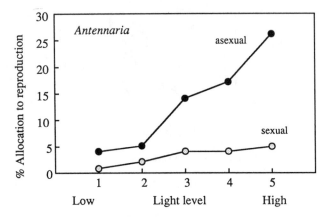

Fig. 3.20. Differences in allocation to reproduction in asexual and sexual populations of *Antennaria* in response to light availability (modified from Michaels and Bazzaz 1986).

as compared to apomicts. Among the sexuals, males and females did differ in response. For example, males consistently had lower mortality and higher allocation to ramet production (Michaels and Bazzaz 1989). The study indicated that apomictic *Antennaria* are well suited for colonization of disturbed sites. They are flexible in response, grow over a wide range of resource gradients, produce more inflorescenses per plant, and are opportunistic in reproduction. Sexual plants, however, showed characteristics that promote competitive ability in later-successional habitats. Sexuals have a greater diversity of individual resource-response curves within populations and a conservative allocation pattern between seed and stolons, whereby a premium is placed on vegetative spread and high adult survival.

There is great interest in clonal growth and a recognition of its importance in succession and its ecological and evolutionary consequences in a wide range of organisms. Nevertheless, the effects of physiological integration on the spread of genets, integration of patch types, and response of clonal shrubs and trees to herbivory have not been fully investigated. Because clonal species commonly dominate mid-successional communities and facilitate recruitment of late-successional trees, they help shape the trends of dominance and diversity in recovering ecosystems. Further study of clonal plants may therefore provide appreciable insight into successional patterns in many ecosystems.

4

The environment of successional plants: disentangling causes and consequences

What creates successional habitats?

Plants respond primarily to changes in their resource base and to the fluxes of chemical and physical factors in their environment. In the process of destroying biomass, disturbances alter the resource base of a site. Viewed from a plant population perspective, therefore, disturbance can be defined as *a sudden change in the resource base in a habitat that is expressed as a readily detectable change in population response* (Bazzaz 1983). The key terms in this definition are 'sudden change' and 'detectable response.' Although these phrases are easily understood, neither can be precisely quantified and, therefore, can be perceived differently by different investigators.

Clearing natural vegetation for agricultural purposes, abandonment of agricultural lands, and disturbance of vegetation by natural causes are the main ways in which successional habitats are created. Some important natural disturbance agents are fires, hurricanes, landslides, earthquakes, volcanic eruptions, herbivores, and pathogens. Each may act either individually or in various combinations (see White 1979, Mooney and Godron 1983, Pickett and White 1985). Because of its prevalence, natural disturbance can be considered an integral component of all landscapes.

Recent human activities especially important in creating disturbed habitats include:

1. Extensive clearing of natural vegetation for agricultural purposes, including plantation forestry and the abandonment of less productive land.
2. Mining activities and use of coal, minerals, oil, and other natural resources. Some of these activities can create scarred and polluted landscapes and can cause acid rain, forest decline, and other damage to ecosystems.

3. Emission into the environment of large quantities of chemicals and other pollutants such as biocides, SO_2, N_2O, CO_2, CH_4, CFCs and other greenhouse gases, some of which can lead to global warming and major vegetational shifts.
4. Creation of war-impacted ecosystems by extensive bombing, defoliation, and movement of soldiers and equipment.

These events are expected to intensify due to increased human population and demands for natural resources. Disturbances of different types can modify resource fluxes in specific ways, especially the degree of heterogeneity in resource availability. For example, craters created by bombs in the rainforests of Southeast Asia harbor aquatic organisms and undergo small aquatic 'hydrarch successions' in otherwise terrestrial environments.

Disturbances of many kinds can be experienced in any ecosystem, but specific disturbance regimes may be more prevalent in certain ecosystems. For example, fire is especially important in grasslands, shrublands, and many forested ecosystems, particularly those with a distinct and prolonged dry season. In many of these habitats, fire plays a critical role for the maintenance of these ecosystems and their biological diversity. Elsewhere, hurricanes can cause extensive disturbance, particularly in high stature temperate and tropical forests (e.g. Foster 1988, Whitmore 1984). Animals are more common disturbance agents in grasslands and shrublands. Pathogens usually exert a patchy influence on vegetation, but their effects can be widespread. The chestnut blight fungus, for example, effectively eliminated chestnut (*Castania dentata*) from its entire range in eastern North America. Herbivory occurs in all natural communities, but episodic herbivory events, such as gypsy moth and spruce budworm infestations, can devastate one or more plant populations over extensive areas, and create patchy successional habitats.

Disturbances occur at several scales, ranging from those sensed by one or a few plants to those perceived by thousands of individuals. Factors that cause disturbance can have different influences on the scale, intensity, and frequency of occurrence of disturbance and on the level of environmental heterogeneity created. These factors can be major selective forces in the evolution of plant life history traits and in determining what species are most likely to occupy certain successional habitats and even specific patches within those habitats. Recently, much attention has been devoted to several global-scale disturbances: the clearing of large tracts of tropical forests (Dixon *et al.* 1994), rapid climatic changes due to increased concentrations of greenhouse gases in the atmosphere (Houghton *et al.*

1990, 1992), and the large-scale effects of war, including several potential 'nuclear winter' scenarios (Ehrlich *et al.* 1984, Sagan and Turco 1990).

Disturbance agents can create distinct successional habitats and greatly influence the heterogeneity of aerial and soil resources. This heterogeneity can influence the spatial distribution of the recruited individuals and the future of the population. Depending on their intensity, fires can consume litter, volatilize much nitrogen and carbon, cause a pulsed release of nutrients, destroy many seeds while enhancing the germination of others, kill soil organisms, and change soil reflectance, energy balance, water-holding capacity, and chemistry. Fires can actually homogenize an initially heterogeneous habitat. Land clearing and plowing for agriculture removes some carbon and mixes the upper layers of the soil profile into a uniform 'plough layer,' which may include a mixture of both organic and mineral horizons. Such mixing improves soil aeration, thus accelerating mineralization and nutrient release. These changes, together with the addition of fertilizers during extended cultivation, can also greatly homogenize the site prior to its colonization by successional plants. In contrast, hurricanes can cause severe defoliation, uprooting, and breakage of trees, which creates a very patchy physical environment, especially with regard to nutrients and light (Walker 1991). They can also create distinct microtopographic features and habitats, and generate patchy distribution of seeds and seedling recruitment. A gap created by uprooting and breakage of trees (in a hurricane) can be a heterogeneous mess.

While it is instructive to search for general trends in the environmental factors of habitats along successional gradients, it is important to recognize the great heterogeneity of environments in these habitats and the importance of this heterogeneity in successional change within patches. Scale is very critical in these analyses.

General trends in the physical environment in successional habitats

Except for some possible differences in reflectivity, there is no *a priori* reason to suspect that early- and late-successional habitats should be inherently different from each other in the amount of radiation received at the surface of the vegetation. Any differences among habitats, therefore, are largely the product of the nature of the vegetation itself and how light quantity and quality change as radiation penetrates the canopy. Factors that can influence the energy microenvironment in successional habitats include the height of the surface of energy exchange between the vegetation and the bulk air, the depth of the canopy, leaf area index and leaf area density, and the thickness of the litter layer above the soil surface. In newly disturbed sites, the region of energy exchange is at or near the soil surface itself. With the invasion of pioneer plants, the region of energy exchange shifts to the top of the canopy and continues to be elevated as the depth of the canopy increases and leaf area index and leaf area density become greater. With the development of greater deciduous leaf area, the litter layer also becomes thicker as succession proceeds. Because of the differences between soil and plants in water content, heat capacity, and conductivity, the microclimate changes dramatically from an exposed site after severe disturbance with bare soil to a mature-phase forest.

The aerial environment

Light

Both light quantity and quality change as the light passes through the canopy. If leaf area is uniformly distributed both vertically and horizontally, then the extinction of light as it passes through the canopy will closely follow the Beer-Lambert formulation $I = I_o e^{-k}$(LAI) where I is the intensity of radiation at a given distance below the top of the canopy, I_o is the radiation incident on the top of the canopy, k is the extinction coefficient for the particular community, and LAI is the cumulative leaf area index above the level at which I is estimated (Larcher 1983). The light intensity profile will be smooth and uniform. Irrespective of its successional status, however, natural vegetation is never uniform, even at a small scale, and therefore the light environment within the canopy is extremely complex (Pearcy 1988). With the continuous changes in the position of leaves with regard to the sun, enormous variability between direct and diffuse light will be encountered both in time and in space. Numerous studies have documented great differences in light intensity among the forest understory, gaps representing early-successional habitats in the mature forest matrix, and large clearings with little vegetation (see Chazdon 1988). Sunflecks of different durations become important for plant carbon gain as ecosystems recover and light levels in the understory are reduced. In late-successional forests sunflecks become critical for the functioning of many individuals in the understory (e.g. Horn 1971). Because of differences among species in crown depth and to a lesser extent in leaf characteristics, various species in an ecosystem cast different shadows and help create further heterogeneity in the light environment (Canham et al. 1994). Furthermore, because of the predominance of chlorophyll (which strongly absorbs red light) over other plant pigments

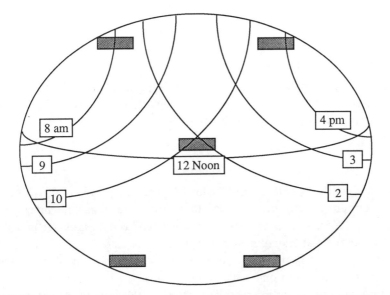

Fig. 4.1. Patterns of direct beam radiation arriving at the forest floor at different hours of the day in July in a temperate forest gap in the Northern Hemisphere (Sipe and Bazzaz unpublished data).

during most of the growing season, the ratio of red/far-red light decreases from the top of the canopy to the ground. The red/far-red ratio in the understory of early-successional fields also changes with the season as the canopy develops. Red/far-red ratios have been shown to influence seed germination and recruitment as well as internode elongation and plant architecture (Salisbury and Ross 1985). Under dense canopies early-successional plants fail to be recruited because of the high far-red/red ratio in these habitats.

The amount of radiation received in gaps of different sizes is predictable with some knowledge of gap geometry and sun tracks (see Canham *et al.* 1990). In temperate forests, for example, northwest sites receive significantly more light than do southwest or southeast sites. Also, the length of exposure of direct beam radiation varies greatly among various locations within a gap (Fig. 4.1). Thus, seedlings in different locations in the gap receive different amounts of total daily radiation, with peak radiation received at different times of the day. In addition to canopy gaps, neighbors can greatly modify the quantity and quality of light received by a plant. Fast-growing species generally have open canopies and so cast less dense shade than do slow-growing species with denser canopies. Effects of neighbors on the light environment can profoundly influence plant growth and development (Pacala *et al.* 1993), and therefore have a significant impact on successional patterns. Other than the intuitively obvious differences among clearings, gaps, and mature forest understory in the total amount of light received near the ground, few other generalizations about the light environment are possible. It can be said, however, that from the perspective of the individual plant (especially in the understory), light is the most variable of all plant resources.

Temperature

Air and soil temperature usually follow patterns of radiation, with soil temperature lagging behind solar radiation more than air temperature (Geiger 1965). Deeper in the soil, the lag time becomes longer and the influence of solar radiation is reduced. Bare soil and vegetation differ in their reflectivity, heat capacity, heat conductance, and amount of heat lost through evaporation and sensible heat transfer. These factors, together with quantity of radiation received, determine the amount of heat storage and thus the temperature of soil and vegetation. Early in the season in early-successional habitats with essentially bare soil, the soil surface is the site of maximum energy exchange with the atmosphere and it therefore experiences the largest daily fluctuation in temperature. With greater canopy development, energy exchange occurs farther from the soil surface and soil temperatures fluctuate less.

Radiation load, transpiration rate, canopy conductance, and energy dissipation by conduction and convection determine the differences between air temperature and canopy temperature. The ability of plants to maintain leaf temperatures that are higher or lower than ambient air temperature may be important for photosynthesis and respiration. High transpiration rates, for example, enable fast-growing early-successional plants on exposed, sunny sites to maintain leaf temperatures several degrees below air temperature. In contrast, winter temperatures of rosettes of many composites and crucifers in temperate, early-successional fields can be several degrees above air temperature. This enables the rosettes to photosynthesize and accumulate carbon even when air temperatures are quite low. Carbon accumulation during winter can correlate positively with plant performance during the following spring and summer. Similarly, in mid-successional fields seedlings of the evergreen *Juniperus virginiana*, which are protected from wind by grasses, can maintain leaf temperatures above ambient air temperature and above leaf temperatures of exposed adults. These seedlings are able to gain carbon and grow during the winter (Ormsbee *et al.* 1976).

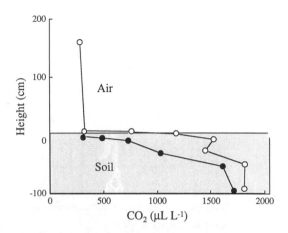

Fig. 4.2. Profiles of CO_2 concentration in soil and adjacent air early (●) in the growing season and late (○) in the growing season in an early successional field (data from Schwartz and Bazzaz 1973).

These examples illustrate that carbon gain and growth in many habitats must be assessed in relation to leaf temperature rather than air temperature.

Carbon dioxide

Despite its critical importance for plant growth as a substrate for photosynthesis, variation in carbon dioxide around plants has not been adequately considered. It was initially assumed that CO_2 is a highly diffusible resource and, therefore, its variability is not large. However, measurements show that it can vary greatly spatially, especially vertically. Carbon dioxide is produced by aboveground plant respiration and by microbial and root respiration from the soil, and is consumed in plant photosynthesis. Depending on the amount of organic matter in the soil and its temperature and moisture, the soil can be the main source of CO_2 for understory plants in the forest during the daylight hours and for seedlings near the ground. In early-successional fields, CO_2 concentration is highest in the soil, especially just above the water table (Fig. 4.2). In the air, the highest CO_2 concentrations occur near the soil surface. During hours of active photosynthesis, concentrations of CO_2 decline within the plant canopy (Table 4.1). Particularly during periods of low air movement, the canopy of fast growing dense stands can become depleted of CO_2, thus causing a complex vertical CO_2 profile to develop. At night, CO_2 concentrations within the canopy are usually higher than they are during

Table 4.1. *Mean* CO_2 *profiles showing values at morning (during the time of maximum photosynthetic activity) and at midnight in an early-successional field in early June. Measurements at 320 cm were situated well above the plant canopy*

Height (cm) above soil surface	Morning	Midnight
320	324	330
80	320	337
20	320	349
5	318	362
−2.5	607	552

Source: Data from Schwartz and Bazzaz 1973.

the day. The highest CO_2 concentrations usually occur near the ground at night, when soil temperatures reach their maximum. Thus shorter seedlings, or lower parts of larger seedlings, can be in a high CO_2 environment, which can greatly influence their total carbon gain (Bazzaz 1974, 1984*b*).

Concentrations of CO_2 can differ greatly in the canopy within and between the seasons. In a secondary forest with low soil organic matter in New England, we measured CO_2 profiles at several heights for a period spanning the active and inactive parts of the year (Fig. 4.3; Bazzaz and Williams 1991). The measurements show both the vertical gradient and seasonal differences in CO_2 concentration. Because of differences among patches in organic matter content, soil moisture, temperature, and exposure, there can also be spatial patchiness in CO_2 concentrations. Significant differences were found among various sites in a temperate and a tropical forest, and these differences became especially pronounced at night. Although early- and late-successional habitats apparently differ in the steepness of vertical CO_2 and light profiles, seedlings in both kinds of habitat can experience higher levels of CO_2 than their parents. However, those in late-successional habitats grow in a less well-lit environment than do those in early-successional open habitats. Thus, for early-successional seedlings, the two environmental factors conducive for fast growth, light and CO_2, are more congruent than for the late-successional seedlings. But in both habitats, smaller (i.e. shorter) seedlings experience high CO_2 and low light conditions, and larger seedlings and saplings experience relatively low CO_2 and high light. Furthermore, as is the case for other environmental factors, different leaves of the same mature individual may be located in microenvironments that differ greatly in CO_2 concentration. This would be

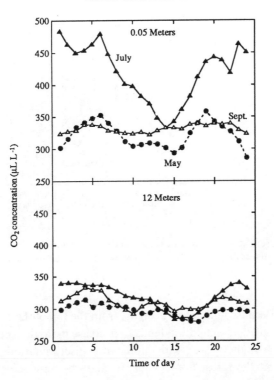

Fig. 4.3. Changes in CO_2 concentration measured at two heights during the growing season in a temperate deciduous forest (modified from Bazzaz and Williams 1991).

more pronounced for seedlings of late-successional plants and other individuals near the forest floor in the understory. Mixing by wind in open early-successional habitats may reduce this effect.

Seedlings apparently use a substantial amount of CO_2 that has been respired by plants and soil organisms as substrate for photosynthesis. Sources of CO_2 are being elucidated by stable isotope techniques (e.g. Medina *et al.* 1991, Ehleringer *et al.* 1993, Farquhar *et al.* 1993). Being mostly a product of soil respiration, carbon dioxide near the ground is isotopically light (low $\delta^{13}C$). Because of stomatal and *Rubisco* discrimination against ^{13}C, seedlings exposed to CO_2 fluxing from the soil will differ in isotopic composition from canopy foliage, which is exposed to both recycled CO_2 and CO_2 in bulk air which has a higher ^{13}C content. Furthermore, seedlings in the understory grow in lower light levels, and have a higher ratio of intercellular to ambient CO_2 concentration (C_i/C_a). These factors influence the discrimination between ^{13}C and ^{12}C (Farquhar

Environment of successional plants

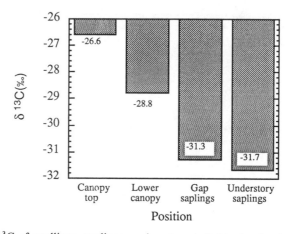

Fig. 4.4. $\delta^{13}C$ of seedlings, saplings, and mature individuals of red maple (*Acer rubrum*) in a second growth forest (Bassow and Bazzaz unpublished data).

et al. 1982). Because light levels decrease from the top of the canopy to the forest floor, carbon assimilation tends to decrease more than conductance, resulting in a higher C_i/C_a ratio and more discrimination by *Rubisco* against ^{13}C at the forest floor. Thus, in mature-phase ecosystems with relatively high soil respiration, leaves of seedlings should be isotopically lighter than leaves of their parents (Medina *et al.* 1991). Gaps that disrupt the canopy and mix air would reduce the isotopic difference between juveniles and adults (Fig. 4.4).

Despite its great importance in succession, the soil environment has received much less attention than the aerial environment in studies of plant response. Soils in early-successional habitats, created after severe disturbances or repeated cultivation, can differ dramatically from soils of late-successional habitats. Agriculturally abandoned early-successional soils may have undifferentiated upper horizons mixed together by repeated discing and plowing. In sites with topographic relief, erosion can be severe and may lead to the removal of the top layers of soil from higher areas and their deposition on lower parts of the field. In such situations erosion creates more uniformity in topography but greater heterogeneity in soil texture and organic matter content, and therefore in nutrient and water availability. Plowing increases soil aeration and therefore oxidation, and can greatly influence microbial activities, mineralization rates, and nutrient availability. Soil properties of early-successional habitats, especially their nutrient availability, will depend on the nature of disturbance. In fallow agricultural fields, soils will also reflect the kind and amounts of plant residues and residual fertilizers and biocides used during cultivation.

Nutrient availability and dynamics

Succession is most commonly studied on inherently poor or degraded soils that have been cleared for agricultural purposes. Whatever the initial soil properties before clearing, soils *do* change during ecosystem recovery, both under the direct influence of the changing vegetation and indirectly through the interaction of vegetation with the local climate. The most dramatic changes occur during the first few years of succession, especially in the upper portions of the soil profile. The addition of organic matter, the development of a litter layer and the stabilization of the soil surface are the major factors in soil changes with succession. Because of the great differences in initial conditions, trends in soil properties with succession can vary greatly on different sites. Inherent nutrient availability controls the trends in nutrient relations with succession. Availability of nitrogen, and to a lesser extent phosphorus, is critical for ecosystem recovery. Therefore, many ecosystem restoration programs involve nitrogen and phosphorus amendation (see Bradshaw and Chadwick 1980).

Patterns of nitrogen availability in succession have been studied in only a few places. Nitrogen supply to plants is expected to increase greatly following disturbance, and then may continue at low levels for several years afterwards (Matson and Vitousek 1981, Robertson and Vitousek 1981, Robertson 1984; see Fig. 4.5). In inherently rich soil and in soils that receive heavy doses of fertilizer during cultivation, nitrogen availability can decline during the first few years of succession. In contrast, on sandy nutrient-poor soil nitrogen concentration generally increases with succession (Inouye *et al.* 1987, Tilman 1987). On these soils, nitrogen mineralization rates increase with total nitrogen content and therefore also increase with succession (Pastor *et al.* 1987). Thus, trends of nitrogen availability in secondary succession on sandy soil may be similar to those of primary succession on sand dunes, as shown in J. Olson's classic study (1958). Trends in other elements are less well understood, but undoubtedly there are differences between early- and mid-successional fields. Some studies have shown a decline during the first few years of succession in soil organic matter content, organic carbon, soil pH, and total nitrogen, magnesium, calcium, potassium, phosphorus, and other nitrogen. However, there are presently no known clear-cut trends in many of these parameters. Initial soil characteristics such as fertility, temperature, and moisture are critical

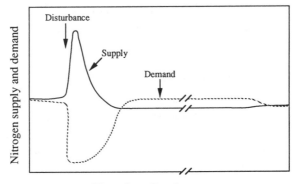

Time since disturbance

Fig. 4.5. Change in supply and demand of nitrogen during recovery following disturbance in forest ecosystems (modified from Vitousek and Walker 1987).

determinants of nutrient dynamics during succession. Because these soil characteristics vary both within and between regions, we might expect few general trends in nutrient dynamics.

Nutrient dynamics in ecosystems are governed by the balance between supply from the soil and demand by the plants. When the vegetative cover is low, as it is in early phases of recovery, the supply of nitrogen and other nutrients in many soils may exceed the demand of the plant's biomass. In such circumstances, the system can export large quantities of mobile nutrients such as nitrate (e.g. Bormann and Likens 1967). As the system accumulates biomass, however, the demand for nutrients can exceed the supply from the soil, resulting in negligible transport of nutrients out of the system (Vitousek and Reiners 1975, Vitousek and Walker 1987). In many ecosystems, demand and supply reach a dynamic equilibrium in late succession (Vitousek and Reiners 1975, Bormann and Likens 1979). Recently, however, large quantities of nitrogen, primarily in the form of NO_3, are being deposited on many ecosystems through human activities (e.g. Lovett and Kinsman 1990). These additional nitrogen inputs, together with *in situ* mineralization, can cause nitrogen supply to exceed the demand by plants. In such cases, the system exports nitrates to adjacent ecosystems or N_2O to the atmosphere or is poisoned by excess nitrogen (Aber *et al.* 1989, Schulze 1989). The nitrogen economy is being altered in both early- and late-successional habitats so that some previously nitrogen-limited ecosystems are becoming nitrogen saturated (Vitousek 1994). Nitrogen saturation is becoming an important factor in ecosystem recovery in highly polluted regions. Changes in nitrogen supply are expected to influence the rate of successional changes and may also influence their trajectories.

The forms of nitrogen

The two forms of nitrogen commonly used by plants are nitrate (NO_3^-) and ammonia (NH_4^+). These forms differ from each other in mobility and in cost of assimilation by the plant. Nitrate is more mobile, easily leached, and must be reduced (with the aid of the enzyme nitrate reductase) before it can be incorporated into plants. In contrast, ammonia is less mobile, less leachable, and can be readily incorporated into nitrogen-containing compounds in the plant. Soils differ in the ratio of NO_3^- to NH_4^+, and this ratio can change during ecosystem recovery. There is usually an increase in the NH_4/NO_3 ratio during the recovery of forest ecosystems (Likens *et al.* 1970) and in oldfield succession. In oldfields, this relative increase in NH_4 availability is particularly pronounced during the time of dominance of species such as *Solidago* (Burton and Bazzaz 1995), which produce allelochemicals that inhibit nitrogen-fixing bacteria (Rice 1984, Rice and Pancholy 1972, also see Vitousek and Walker 1987). The increase in nitrogen deposition and associated shifts in the NO_3/NH_4 ratio can differentially influence species and modify successional trajectories. For example, nitrogen deposition can alter the light responsiveness and shade tolerance of seedlings (Bazzaz *et al.* 1990).

To detect differences among species in response to changes in the nitrogen economy, we examined the effects of simulated nitrogen-deposition treatments on four successional birch species that are present in New England forests. *Betula populifolia* and *Betula papyrifera* are early-successional species; *Betula lenta* and *Betula alleghaniensis* are late-successional species. Nitrogen was applied as either NO_3^-, NH_4NO_3, or NH_4^+. *Betula lenta* showed a significant growth response to nitrogen form, especially at low light. Under low-light conditions, NO_3-supplied plants grew best, perhaps because of greater allocation to leaf tissue and greater specific leaf area (Crabtree and Bazzaz 1993*a*). Under high-light conditions, NH_4-supplied plants grew best. Since *B. lenta* and *B. alleghaniensis* seedlings persist in the understory, a different species mix of seedlings may be released when a gap is created under changed nitrogen conditions, with potential consequences for competitive interactions and ascendance to the canopy (Crabtree and Bazzaz 1993*b*) (Fig. 4.6).

Nitrogen availability can also greatly influence the response of plants to other environmental resources. There can be interaction between nitrogen

Environment of successional plants

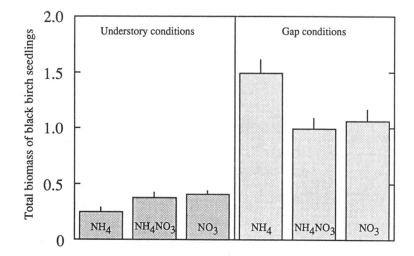

Fig. 4.6. Difference in growth among black birch seedlings (*Betula lenta*) given different forms of nitrogen in high and low light (modified from Crabtree and Bazzaz 1993*a*).

supply and CO_2 levels, which are both increasing in many ecosystems. Differential enhancement of response among species may modify successional trends. For example, the growth enhancement by elevated CO_2 in seedlings of early-successional trees, such as gray birch, is greater at a higher nitrogen input, whereas for relatively later-successional species, such as red oak and striped maple, CO_2 enhancement is most evident under low light conditions (Bazzaz and Miao 1993; Fig. 4.7). Altered nitrogen deposition in terms of quantity and form will likely exert a novel impact on ecosystem recovery.

Soil microorganisms

The role of soil microorganisms in the nutrient dynamics of succession has not been extensively investigated, but microbes are known to be major competitors with plants for soil nitrogen (Melillo and Gosz 1983, Vitousek and Matson 1985). Competition between the microbial community and plants is especially intense in poor soils. Thus, immobilization of nutrients by microorganisms may be a significant component of nutrient dynamics in ecosystem recovery. Seasonal and spatial variation in immobilization of nutrients during succession, and the relationship between immobilization patterns and species response, competitive interactions, and species replacement have not yet been investigated. Also, the succession of

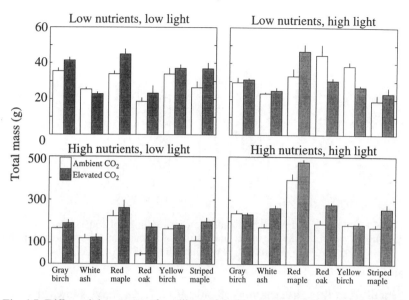

Fig. 4.7. Differential response of seedlings of deciduous trees to light, nutrient, and CO_2 concentration during growth representing understory and gap conditions now and in the future (modified from Bazzaz and Miao 1993).

microorganisms themselves as vegetation changes has not yet been explored.

Mycorrhizal associations, which seem to be common in both early- and late-successional plants, may play an important role in nutrient and water uptake. Mycorrhizae are known to greatly expand the foraging domain of roots and therefore allow them further access to less mobile resources such as phosphorus. The potential for reinoculation of human-disturbed arid land with mycorrhizae is significantly lower than that of adjacent non-disturbed areas (Allen 1991). In particular, intensive disturbances such as strip mining for coal tend to eliminate or severely reduce mycorrhizal activity. Plants that naturally invade these highly disturbed areas are usually non-mycorrhizal. However, in sites where the topsoil is replaced and ecosystem restoration is practiced, mycorrhizae are more common (Allen and Allen 1990). Similarly, Janos (1980) suggests that mycorrhizal activity in tropical forest soil may be virtually eliminated following disturbance and that mycorrhizal activity increases through succession. There is also evidence that species of mycorrhizal fungi shift with the progress of succession (Anderson et al. 1984) and their abundance changes between seasons and years (Allen and Allen 1990).

The few specific studies of the response of plants to mycorrhizal association suggest that even some early-successional plants can greatly

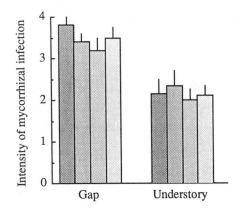

Fig. 4.8. Intensity of mycorrhizal infection in seedlings of four species of birch growing in a gap and in the intact understory of a deciduous forest (unpublished data of Wayne and Bazzaz).

benefit from these associations. Koide *et al.* (1994) have shown that the mycorrhizae greatly increase the growth of the annual *Abutilon theophrasti*, resulting in significantly greater recruitment in the following year. We found that ninety percent of the seedlings of four coexisting birch species that differ in shade tolerance and successional status had ectomycorrhizal infection five months after transplanting into a gap-understory gradient in a mixed deciduous forest. Roots were extensively infected in the exposed north sides of gaps and least infected in the forest understory (Bazzaz and Wayne 1994; Fig. 4.8). In the birches, there was a strong positive correlation between seedling growth and the degree of mycorrhizal infection. However, despite significant differences in growth among the birch species within the gap environment, the differences among them in mycorrhizal infection were not large. Because mycorrhizal infection and growth enhancement interact reciprocally, it is difficult to assign cause and effect. Despite this limitation, understanding natural succession of soil microorganisms and mycorrhizae in different systems should greatly inform ecosystem restoration and management.

Herbivores and succession

Although it is known that herbivores can play a major role in plant community organization (e.g. Harper 1969, Janzen 1981, MacMahon 1981, Crawley 1983, Dirzo 1984, Fritz and Simms 1992), only limited work has been done on their role in recovering ecosystems. Slugs, small mammals, rabbits, and insects are common herbivores in early-successional communities, whereas large herbivores and insects are important grazers in taller, later-successional vegetation. Selective herbivores can influence ecosystem recovery by differentially affecting species in a community. In general, plants are assumed to have evolved defenses against herbivores that they have encountered through their evolution. The effectiveness of these defenses, however, is sometimes reduced through counter-adaptations in the herbivores. This interaction between plants and their specific herbivores is often called the 'evolutionary arms race.' It is expected that with the increase in structural diversity during ecosystem recovery, there will be an increase in the number of herbivores and their predators (Godfray 1985). As with pathogens, however, herbivore importance can vary greatly from location to location and year to year.

The role of herbivores in succession in the oldfields of North America is virtually unknown (Hendrix *et al.* 1988*a*). In Europe, however, there is much evidence that herbivores influence the rate and direction of succession (e.g. Brown 1985). Edwards and Gillman (1987) concluded that composition of the three successional communities that they considered, reflected their history of herbivory. In the early- and mid-successional communities, highly polyphagous insects were important, whereas in the late-successional forest community there was a higher degree of specialization between plants and insects. Herbivory seems to be more important in secondary than in primary succession. Brown and Southwood (1987) suggest that the proportion of herbivorous species of birds and mammals declines with succession, but the proportion and the density of insectivores rises. They have also shown that the herbivore guild structure changes with succession. For example, while the density of phytophagous insects rises, their contribution to the total insect load declines.

Comparing vegetation structure and herbivore impact on succession of a fertile site in Iowa with that of a less fertile site in Silwood Park in England has been quite instructive. Both sites are rapidly colonized by several guilds of insects, with the phytophagous guild being most important (Hendrix, *et al.* 1988*b*). There were fewer phytophages in the structurally more complex New World site than in the English site. The authors attribute the differences between the two sites to dominance by introduced plant species at the Iowa site. Introduced species generally, but not always, carry lower herbivore loads than native plants (Fig. 4.9). At the Silwood Park site, the exclusion of insects by the use of herbicides during the first year of succession significantly increased richness, diversity, and cover of plant species, especially of annuals. In the second year, the exclusion of

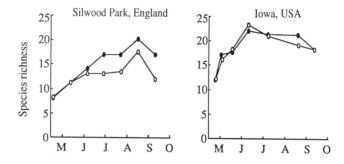

Fig. 4.9. The response of plant species richness to herbivory in Silwood Park, England (left), and in Iowa, USA (right). Insecticide treated (●), control (○) (modified from Hendrix *et al.* 1988*b*).

herbivores greatly increased the abundance of perennial grasses that seem to be particularly susceptible to herbivore damage (Hendrix, *et al.* 1988*b*). At the Iowa site, however, the exclusion of herbivores had little effect on community composition and did not alter the rate or the direction of succession. This is perhaps due to the lesser contribution of susceptible annuals to the vegetation in the first year and of perennial grasses in the second year of succession. The authors suggest that herbivores influence succession in the Iowa sites only when perennial grasses become abundant in the fields later in succession. This observation, however, may not be generalizable. For example, while growing annual plants from successional oldfields of the United States with annual plants of disturbed ground in England in the same glasshouse in Cambridge, England, I observed a severe infestation of the United States annuals by aphids. In contrast, the English species were completely free of aphids.

Causes and consequences of environmental variability and heterogeneity

Due to the influence of theories of evolutionary biology (e.g. Levins 1968), plant ecologists have not only considered changes in environmental predictability and heterogeneity in terms of ecosystem recovery, but also their role in the evolution of plant life history in general. However, some fundamental questions about the level of variability and heterogeneity and how they change with succession remain unanswered. These questions are intricately connected to the emerging issues of scale and hierarchy (Allen and Starr 1982, O'Neill *et al.* 1986, Kolasa and Pickett 1991, Ehleringer and Field 1993, Levin 1993). We still need more definitive answers to the following fundamental questions about variability and ecosystem recovery:

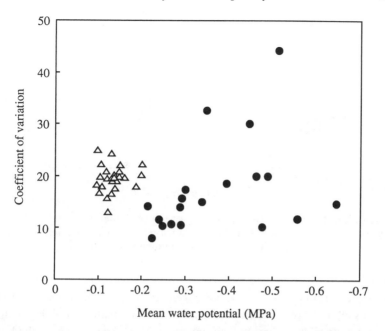

Fig. 4.10. Variation in soil–water potential in an early-successional field (●) and adjacent mature-phase deciduous forest (△) (Rice and Bazzaz unpublished data).

How do we measure variability across spatial and temporal scales?
How are environmental variability and predictability 'perceived' by the plant?
Are environmental factors in a given site equally variable? If not, how do various plant traits evolve to cope simultaneously with variable levels of some factors and more or less constant levels of other factors?
From the plant's perspective, are early-successional environments more variable than late-successional ones?

To compare the general differences in heterogeneity between early- and late-successional communities, we quantified both between-location (coarse grain) and within-location (fine grain) variability in photosynthetically active radiation (PAR) and soil moisture conditions using an early-successional field, a late-successional grassland, and a mature deciduous forest, which were all located adjacent to each other. We found that large-scale spatial variance in PAR was lower on the forest floor than in either the field or the grassland, where it did not differ. The total variance in soil–water potential decreased from field to grassland to forest (Fig. 4.10).

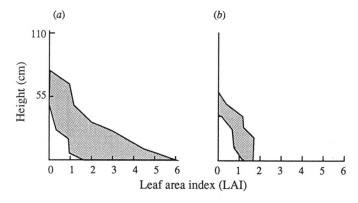

Fig. 4.11. Difference in leaf area profiles, showing maximum and minimum leaf area measured in an early successional (*a*) and adjacent mature phase forest understory (*b*) (Rice and Bazzaz unpublished data).

The habitats also differed in the distribution of leaf-area index, which is a strong determinant of vertical heterogeneity in light conditions (Fig 4.11). Plants in the early-successional field experienced greater variability in light than plants in the other two communities. These conditions resulted in a more complex environment in the early-successional field. Furthermore, there was a temporal component to this variability, as soil moisture variability and differences in variability among habitats were greatest during periods of drought and generally increased as the season progressed.

   Whatever the relative level of heterogeneity may be, the microenvironment of successional habitats is a dynamic mosaic varying in time and space. The vegetation itself is usually a mosaic of patches which may be dominated by species that differ in identity, stature, architecture, physiology, nutrient requirements, etc. In fact, because of the nature of invasion and recruitment and the co-presence of different plant forms, mid-successional habitats in a given region may be the most patchy of any successional vegetation. For example, even in a small area (10 ha) on a level terrain we can identify several distinct mid-successional patch-types (Burton and Bazzaz 1995) that differ greatly in their microenvironment and in their hospitality to various invading trees. Furthermore, an analysis of 21 site and vegetation characters indicates that differences among distinct successional patches can change sharply in space and time. Identification of patch type in vegetation by observation can be subjective and complex. However, multivariate techniques using measured attributes can simplify this complexity and produce quantitative classifications of patch types.

   It is clear that there are no plants without an environment and that the environment of a plant is influenced by the plant's own activity. Therefore, separating these two dialectical components to study their relationship is conceptually and physically difficult. Studies of plant–environment interactions implicitly recognize this intimate coupling and attempt to consider their reciprocal interactions.

In summary:

1. Irrespective of successional status or habitat type, the environment of a plant is complex and dynamic.
2. The environment is influenced by the plants themselves and they react to it: a dialectic.
3. Different environmental factors in a given location may change at different time scales.
4. Various parts of the same individual may experience different environments.
5. Plants must cope with much spatial and temporal variation in the environment, perhaps more in early- than in late-succession.

5
Recruitment in successional habitats: general trends and specific differences

Seed and seedling bank dynamics

The seed bank of a site is made up of seeds of a variety of plant species dispersed from adjacent sites or dropped locally by the few remaining mature individuals present at the site. Much colonization of forest sites occurs from seedlings and saplings that are already present in the site and from sprouting and broken individuals. In abandoned agricultural fields of temperate regions, the seed bank contributes most of the summer annual plants because their seeds remain viable for several years in soil. The winter annual crucifers are also recruited mostly from the seed bank. In contrast, most of the winter annual composites, although they can be represented in the seed bank, are recruited from newly dispersed seeds. Generally, the contribution of the seed bank to the development of the community diminishes with succession, and the seedling bank becomes the major contributor in mature-phase ecosystems. Regardless of the stage of recovery, seeds of many plant species, including those of late-successional trees, can be dispersed at any time into successional habitats depending on time of maturity and proximity of seed source. The initial stages of dispersal during recovery are aided by wind; later, other agents (especially birds, mammals, and ants) become important dispersal agents.

The seed bank is in a dynamic state of flux. New seeds are added, some seeds germinate, some become deeply buried in the soil, and some die. Persistence in the seed bank differs greatly among species. Although the patterns of decay of seeds of successional plants in the soil has not been adequately investigated, seeds of many summer annuals are known to persist in the seed bank for many years. Thus, when the soil of late-successional habitats is disturbed, the site rapidly is dominated by early-successional plants recruited from this seed bank. While this seems to be the case for

82

many species in temperate and tropical succession, there can be exceptions. Recent work (Alvarez-Buylla and Martínez-Ramos 1990) suggests that seeds of *Cecropia*, a very common and widespread early-successional tree in the humid tropics, do not survive for more than a year. *Cecropia* seeds are rather small, and this fast turnover in the seed bank may reflect seed size (see Vázquez-Yánes and Orozco-Segovia 1993). However, the relationship between seed size and persistence in the seed bank is not well established. For example, irrespective of their size, seeds of many tropical forest species germinate soon after dispersal, forming seedling banks. Those that fail to germinate are quickly eaten by seed predators (e.g. Whitmore 1984). The seedling bank, rather than the seed bank, is thus the major source of recruitment in these tropical forests.

The role of the seed bank in succession has been considered by several authors (see Leck *et al.* 1989). However, because of the enormous difficulty in excavating seeds and measuring seed bank dynamics, there is presently little detailed information on the spatial and temporal dynamics of seed banks in recovering ecosystems. Seeds are very unlikely to be uniformly distributed spatially in the field because of the interaction of several factors, including the location and fecundity of the parents, the heterogeneity of the physical environment (including microtopography), and the patterns of seed dispersal, seed predation, and decay. An indication of the patchiness of seed distribution is found in samples taken from adjacent forest-derived and grassland-derived soils in a single early-successional field. The two soil types, treated uniformly for over 60 years of cultivation, contained very different numbers of germinable seeds (2047 vs. 511 per square meter). The number of recruits in the field usually reflects the size and composition of the seed bank. On forest soil, the peak density of plants was about 400 plants m^{-2}, and on the grassland soil, it was only 160 plants m^{-2} (Fig. 5.1).

Prior to recruitment, and for some time afterwards, the soil surface is the place of exchange of energy with the atmosphere. It is near the soil surface that the greatest discontinuity occurs between the two phases that plants experience in their surroundings, the solid phase and gas phase. Because of the great disparity between soil and air in heat capacity and heat conduction, maximum fluctuations in temperature occur at the surface (Fig. 5.2). Because of intense reradiation from the exposed soil surface, its temperature can be quite low at night. The seeds that are being recruited and the young seedlings that live in this temperature regime must cope with this great variation in temperature. Furthermore, various depths within the soil differ greatly in their microenvironments, especially in the degree of temperature fluctuations and CO_2 concentrations. As environmental

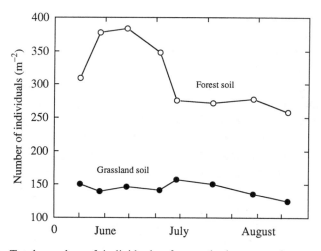

Fig. 5.1. Total number of individuals of annual plants growing on adjacent grassland-derived and forest-derived soils in the same field (modified from Raynal and Bazzaz 1973).

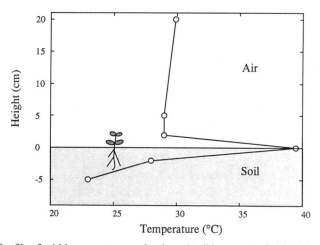

Fig. 5.2. Profile of midday temperature in air and soil in an open field in May during the time of seedling recruitment in the field. Young seedlings in early-successional fields have to endure great fluctuations in temperature between day and night.

gradients can be extremely steep in soil, a change of a few centimeters in depth can produce a very different thermal environment. The deeper the seed is located in the soil, the lower the mean temperature and the narrower the daily fluctuations it experiences at the time of recruitment, and the more

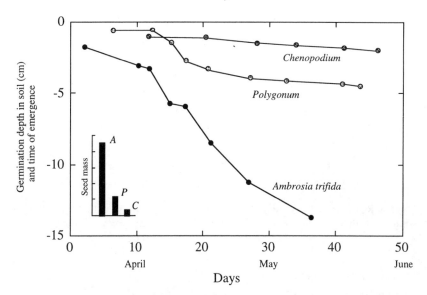

Fig. 5.3. Seed mass, depth of germination, time of emergence of the annuals *Ambrosia trifida*, *Polygonum pensylvanicum*, and *Chenopodium album* in the field during late March, April, and May.

energetically costly it is for the seedling to emerge above the soil surface.

In infrequently disturbed ecosystems, seed density in the soil declines with depth. However, plowing and discing during cultivation can mix seeds such that they are placed more or less uniformly in the upper 15–20 cm of the soil. Therefore, after abandonment, recruitment can occur from any depth in these highly mixed plow-layers. However, only large-seeded plants can be recruited from lower depths (Fig 5.3). Seedlings derived from small seeds are likely to expire before they reach the soil surface. Therefore, recruitment is greatly influenced by location of seeds in the soil through the interaction of seed mass and the response of seeds to microenvironmental fluctuations. Irrespective of seed size or genetic identity, deeper germinating seedlings reach the soil surface later and are, therefore, likely to become subordinate in the canopy. Community structure and size hierarchies and the level of genetic variation in local populations can be greatly influenced by these interactions. For example, seed germination of the relatively large-seeded early-successional annual *Ambrosia trifida* can occur in the field at a wide range of soil depths. Therefore emergence above the soil surface can occur over several weeks. Because their earlier-recruited neighbors are already growing, and because these seedlings have lost energy by emerging from deep in the soil, these late recruits are competitively

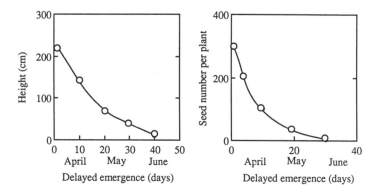

Fig. 5.4. Reduction in final plant height and seed number of individuals of *Ambrosia trifida* recruited in the field at different times in early spring (modified from Abul-Fatih and Bazzaz 1980).

inferior and can suffer much mortality. The survivors are much shorter, smaller, and produce many fewer seeds than the early recruits, especially in dense populations where plant–plant interactions begin early in life (Fig. 5.4). In contrast, in sparse populations delayed recruitment in this species does not carry much cost in terms of performance.

The contribution of the seed bank is less important for invading trees than it is for early-successional herbs. Except for a few early-successional shrubs and trees such as *Rubus* (Whitney 1984) and *Prunus* (Marks and Mohler 1985), seeds of invading woody species do not usually survive in the soil for a long time. Thus, the distance to seed source and differential efficiency among species for wind dispersal dictate the identity, number, and spatial distribution of early-successional tree recruits. Later in succession, the availability of attractive rewards will encourage animals to frequent those sites and transport seed. Finally, the spatial distribution of the early tree colonists, when they have grown above the herbaceous layer, becomes important in recruitment, as they can simultaneously serve as a seed source and as sites for bird and other animal-disperser activities.

Seed germination

Early- and late-successional plants differ in their germination requirements (see Fenner 1985). The majority of early-successional plants, like many other species that thrive in open, sunny habitats, require light for germination. They are usually inhibited by light passing through vegetation, which becomes depleted of red light and therefore has a lower red/far-red ratio.

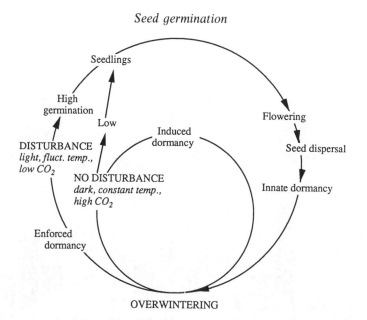

Fig. 5.5. The germination cycle of *Ambrosia artemisiifolia* showing the interaction of dormancy and disturbance in recruitment of populations in early-successional fields (modified from Bazzaz 1979).

Many early-successional species require fluctuating temperatures and are inhibited by high CO_2 concentrations in the soil. Some seeds in temperate environments go into a 'secondary dormancy' if they are not exposed to the appropriate conditions for germination. They must then go through another wintering cycle before they become ready to germinate. Their germination seems to have evolved such that it is highly sensitive to soil disturbance and the removal of vegetation. *Ambrosia artemisiifolia*, the archetypal early-successional annual, has a complex germination behavior and clearly illustrates the strong relationship between recruitment, seed physiology, and soil disturbance in the spring (Fig. 5.5). *Ambrosia* seeds live for many years in the soil. Their germination is sensitive to light, fluctuating temperature, and CO_2 concentration. Disturbance of the soil brings more seeds near the surface and destroys the canopy, thus creating the fluctuating environmental conditions conducive to copious germination. Patterns of development of secondary seed dormancy and enforced dormancy interact with soil temperature to produce changes in the relative dominance of species of the early-successional community. Therefore, when soil disturbances occur at different times during the growing season, there is a clear shift in dominance among species in the early-successional community (Fig. 5.6). And because various species have different impacts on invading plants,

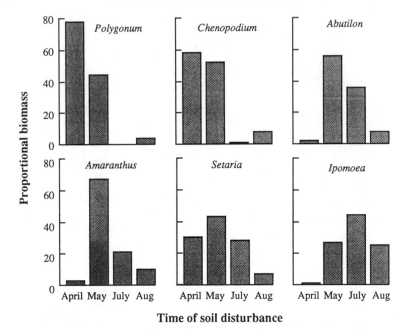

Fig. 5.6. Change in species importance in relation to time of soil disturbance in an early-successional community (modified from Perozzi and Bazzaz 1978).

initial patterns of colonization will shape the structure of the communities that develop later on these sites. The role of such 'initial conditions,' including identity and spatial distribution of early recruits, on the future pattern of community structure is yet to be adequately explored.

In temperate early-successional fields, most seeds of the summer annuals germinate in a relatively short period early in the spring. In these fields the soils are usually moist during the time of seed germination. The species have similar germination patterns along moisture gradients (Fig. 5.7), but differ greatly in their response to temperature (Fig. 5.8). Because of this overriding influence of temperature, the germination of these plants can be successfully modeled as an Arrhenius process (Goloff and Bazzaz 1975). In contrast, winter annuals usually germinate during an extended period in late summer and early autumn, when soil moisture can be limiting. The germination of these plants is thus controlled mostly by soil moisture (Fig 5.9). The pattern of temperature increase in the spring can interact with depth of germination to generate specific seedling populations and dictate the structure of the summer annual communities, especially early in the

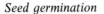

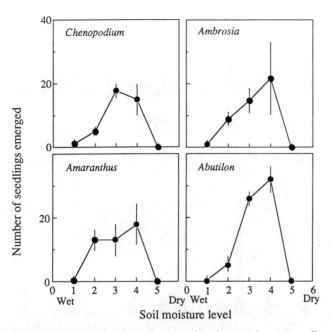

Fig. 5.7. Seedling emergence of early-successional annuals on a soil moisture gradient (modified from Pickett and Bazzaz 1978*b*).

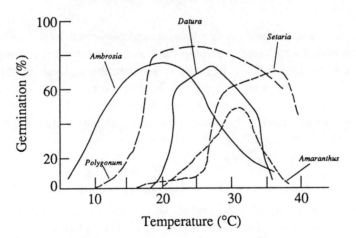

Fig. 5.8. Patterns of germination of five dominant early-successional annuals on a controlled temperature gradient (modified from Bazzaz 1984*a*).

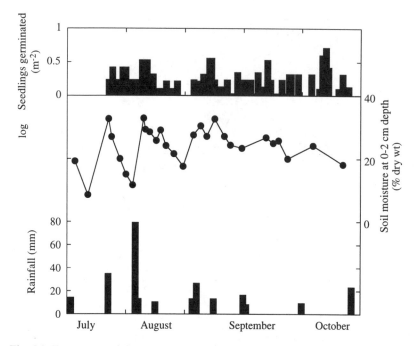

Fig. 5.9. Emergence of the early-successional winter annual *Erigeron canadensis* is modulated by rainfall and soil moisture, and can occur throughout the summer and fall (data from Regehr and Bazzaz 1979).

season. These interactions can generate patch-to-patch and year-to-year differences in community composition. It is noteworthy that in temperate successions differences in response to temperature and moisture among successional tree species seem to be less pronounced than among early-successional herbaceous plants. Again, species appear to be more similar to each other in their response to moisture than in their response to temperature (Fig. 5.10).

Recruitment of clonal plants in successional habitats

Populations of clonal early- and mid-successional plants usually develop mainly from early recruits. Except during the early phase of invasion, the great majority of seedlings die soon after germination. Thus, for many clonal plants in succession, population growth is largely the result of clonal expansion through the production of new ramets from the early-established genets. Even for species that persist and copiously reproduce for several years in the same location, such as goldenrods (*Solidago*), recruitment from

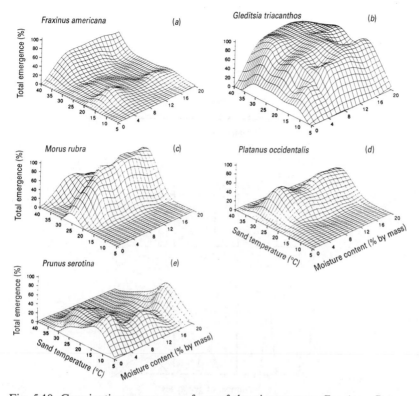

Fig. 5.10. Germination response surfaces of the pioneer trees *Fraxinus, Prunus, Morus, Gleditsia,* and *Platanus* on temperature and moisture gradients (modified from Burton and Bazzaz 1991).

seed is important only during the initial colonization, and the species spreads in the successional field largely by clonal growth. This pattern of recruitment results in many fewer genets than ramets in the population. Because recruitment of germinants both from seeds produced locally and those that arrive from other fields is extremely small, the long-term genetic diversity of the population is determined mainly by the identity of the early recruits. For the composite perennial herbs common in many mid-successions in the temperate zone, the combination of sexual reproduction (which results in the production of a very large number of small, highly dispersible seeds) for colonization and clonal expansion ensure that they arrive in open fields and quickly expand to dominate them. Some clonal shrubs and trees such as *Sassafras* and *Diospyros,* which are initiated from root fragments during cultivation, form extensive thickets under which late-successional tree seedlings can be found. Other common and widespread clonal plants

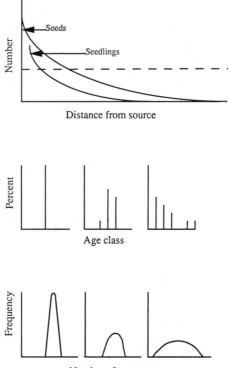

Fig. 5.11. Hypothetical seed shadows, seedling recruitment curves, age class distribution, and genetic diversity in relation to distance from source and severity of disturbance in forest ecosystems (modified from Bazzaz 1983).

such as *Rhus* can also function as nuclei for invasion of the late-successional trees, but arrive in the fields somewhat later.

Patch dynamics and the recruitment of trees

Factors that control kinds, quantities, and spatial distribution of seeds and seedlings of shrubs and trees have been studied in several successional fields and forest gaps. As mentioned earlier, adjacent landscapes with older vegetation are major sources of seed for early-successional fields and gaps. These mature-phase landscapes can exert much influence on the rate of ecosystem recovery, especially on the identity and the spatial pattern of the late-successional plants. Competitive and genetic neighborhoods determined by these patterns, though not yet explored, can undoubtedly play a major

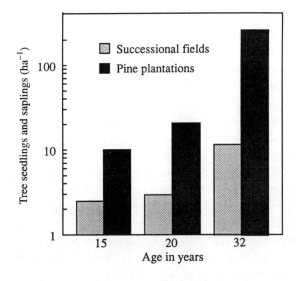

Fig. 5.12. Planting pines in mid-successional fields greatly speeds up succession (facilitation) by encouraging recruitment of a very large number of seedlings of the mature phase forest.

role in future vegetation (Fig. 5.11). Patches of different species may further influence spatial patterns by creating specific microhabitats, which act as filters by differentially encouraging some species and excluding others from establishment.

The spatial pattern of tree distribution can also be influenced by the mode of seed dispersal. Among the early-successional trees in deciduous forests, small-seeded, animal-dispersed species (e.g. *Prunus*, *Gleditsia*, and *Juniperus*) are usually found in scattered patches in oldfields. In contrast, wind-dispersed species, such as *Ulmus*, *Acer*, and *Fraxinus*, can be widely scattered but are particularly common in areas near their parents. Bird-dispersed species generally show much stronger clustering than wind-dispersed species. In fields with exposed soils and less hospitable microenvironments for trees, cover provided by shrubs or early-successional trees greatly facilitates the recruitment of mature-phase trees. For example, where pines were planted in sections of severely eroded oldfields dominated by the grass *Andropogon*, tree seedling recruitment was much greater than in the adjacent open field, and tree seedlings became especially abundant as the pine stands got older (Fig. 5.12). The presence of pines as perches for dispersers, the increase in litter thickness that facilitates germination, and the amelioration of the microenvironment, mostly by shading, play a significant role in enhancing recruitment. These sorts of facilitation of invasion have been documented in several successions (e.g. Debusche *et al.* 1982 in Southern Europe) and have been demonstrated experimentally in New Jersey oldfields (McDonnell and Stiles 1983). Patterns of dispersal and recruitment also greatly influence species richness, as will be shown later.

Mid-successional habitats, such as the pine ecosystem described above, can be the most patchy of all successional habitats. This patchiness is greatly influenced by the patterns of invasion and the recruitment of shrubs and early-successional trees. Clonal spread can also generate patchiness on a smaller scale. In fallow fields, microtopographic features and patterns of erosion, which can become accentuated after field abandonment, can also sort out species distributions along complex gradients (Fig. 5.12). Because of their differential hospitability to various species, highly variable patches can act as 'selective filters' for the recruitment of mid- and late-successional trees. In late-successional habitats, dense understory populations, particularly ferns, seem to act as selective filters in many temperate forests. For example, *Gleditsia triacanthos* emergence was highest in younger patches dominated by *Setaria* or *Aster*, but *Acer saccharum* and *Morus rubra* showed little preference for patch type. Other common, dense understory palms may play similar roles in tropical rain forests.

We studied succession trajectories in an experimental oldfield in the midwestern United States. After 20 years of succession, eight different patches of vegetation could be readily recognized in our field. A study of their microclimate and soil conditions revealed significant differences among them (Burton and Bazzaz 1991). The patches differed especially in temperature and soil moisture. In general, temperature was influenced greatly by relative canopy closure and leaf area index (LAI). Mean temperature, maximum temperature, and range were greatest in *Poa* dominated patches and lowest in *Prunus* dominated ones. Maximum temperatures were very dependent on vegetation type. During the growing season, temperatures exceeded 42 °C in *Poa* patches, but were usually less than 30 °C in *Prunus* patches. Soil moisture was less influenced by patch type than was temperature. Nevertheless, there were differences among them. For example, moisture values were highest in *Prunus* and *Gleditsia* patches (23%) and lowest under *Setaria* (13%). The patches also differed in litter thickness, soil–water potential, root density, subcanopy irradiance, and eight of the ten soil chemistry variables considered. When the positions of the patches were mapped in environmental space using principal component analysis (PCA), the first principal axis explained almost two-thirds of the total variance and was highly correlated with temperature. The second principal axis explained one-fifth of the variance and was strongly associated with litter depth. These factors can change sharply over time, generating a complex dynamic mosaic of microenvironments in the field. Furthermore, these differences can account for the patterns of recruitment and growth of the later-invading species and help determine the spatial pattern of the future community. Emergence, survivorship, growth, and photosynthesis differed among several invading tree species in these different patches. Patch type also appears to be more critical than other habitat factors for seedling emergence. For *Gleditsia*, a three-way ANOVA shows that patch identity accounts for 5% of variance, year for 37%, and soil type for 28%. In general, emergence was negatively associated with LAI, suggesting minimal light requirement for germination and the negative effect of the high far-red/red ratio under dense canopy. The fact that there is higher tree seedling emergence in more or less open sites has also been observed in other oldfields (e.g. Harrison and Werner 1984, Armesto and Pickett 1985, Goldberg and Gross 1988).

Water stress for *Gleditsia* was most pronounced in *Prunus* patches and lowest in *Poa* patches. When measured under the ambient light conditions of the patch, rates of photosynthesis were 9 to 50% lower than rates in the open (without competition) (Burton and Bazzaz 1995). The competition-induced reduction in photosynthesis was greatest for the more shade-intolerant species (Fig. 5.13). Generally, seedling gas exchange rates responded to subcanopy irradiance as well. Height growth was reduced in all patches relative to controls (without competition). Seedlings and saplings of the late-successional species, *Acer saccharum*, were the most competition tolerant, while the early-successional species, *Prunus serotina*, was the least tolerant. *Setaria* patches were most inhibitory to tree seedling performance. Although height growth of tree seedlings differed in various patches and, in particular, between patches and controls, these differences were not dramatic or consistent from year to year. During the first year of growth, the general performance of seedlings in these patches reflected the size of the seed bank, with the large-seeded *Gleditsia* most tolerant of competition. However, in subsequent years, the ranking was independent of seed size, being in the following order: *Acer* > *Crataegus* > *Gleditsia* > *Fraxinus* > *Prunus*. Growth of tree seedlings and saplings was also generally better in grass dominated patches.

Two different kinds of patches are common in oldfield succession and forest succession. In oldfields, patches dominated by different species and life forms, including clumps of shrubs and early-successional trees, can be the sites of recruitment of later-successional trees. They are usually less open and more shaded than the matrix communities. In contrast, gaps and

Recruitment in successional habitats

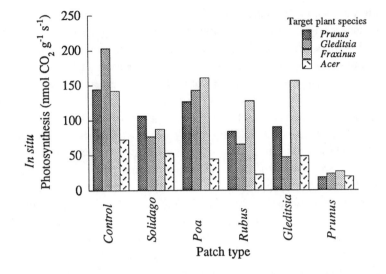

Fig. 5.13. *In situ* photosynthetic rates of *Prunus*, *Gleditsia*, *Fraxinus*, and *Acer* seedlings found in the open with no competing vegetation and under different patches of herbaceous and woody plants scattered in a mid-successional field (modified from Burton and Bazzaz 1995).

other patches created by disturbances in forests are more open than the matrix community. Species of early- and then late-successional trees can be recruited in these patch types. Tree species seem to partition the gap understory gradient, and as the gap closes there is a decrease in early-successional, light demanding species and an increase in shade-tolerant species.

As in all other situations, plant–environment interactions in these patches are reciprocal. The recruited species themselves further modify the patch environment. For example, compared to other species, *Festuca* is very demanding of and removes much phosphorus from the soil. *Solidago*, by producing chemicals that inhibit nitrifying bacteria, leads to an increase in NH_4/NO_3 ratios in patches where it dominates. Species differ in the amount of shade they cast, in their litter quantity, and in their chemical interaction with soil microorganisms that further modify soil chemistry. Thus, site characteristics influence species recruitment and are in turn influenced by them. Despite these differences among patches in their hospitality to the invading trees, these relatively resource-rich sites seem less hospitable than are open sites to tree recruitment (Fig. 5.14). Conversely, however, in highly-eroded fields and on very poor soil, late-successional tree recruitment occurs only in the shade of patches dominated by the shrubs (Bazzaz 1968). Consequently, whether most recruitment of trees

Gleditsia (a)

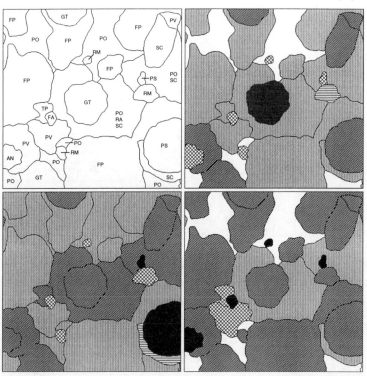

Fraxinus (b)   Prunus (c)

Fig. 5.14. Location of patches of different identities and their degree of resistance to invasion by trees. Key to species composition of patches: AN, *Acer negundo*; FA, *Fraxinus americana*; FP, *Festuca pratensis*; GT, *Gleditsia triacanthos*; PO, *Poa pratensis*; PS, *Prunus serotina*; PV, *Prunus virginiana*; RA, *Rubus alleghaniensis*; RM, *Rosa multiflora*; SC, *Solidago altissima*; TP, *Trifolium pratense* (unpublished data of Burton and Bazzaz).

occurs in the open or under the shade of shrub patches depends on the relative hospitality of those two habitats in a given location dictated by their microenvironment. Different vegetation patches can inhibit or facilitate recruitment by depleting resource levels or by modifying them.

The fate of recruits

Genetic differences among individuals, the nature of the environment at the time of seed maturation (maternal effects), the soil depth from which

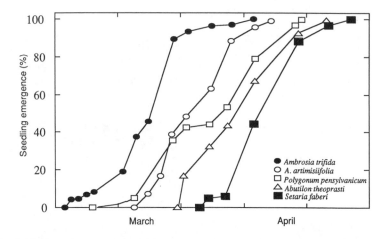

Fig. 5.15. Recruitment times of five major annuals in an early-successional field (modified from Abul-Fatih and Bazzaz 1979*b*).

seedlings emerge, and the nature of the physical environment during germination determine the time of recruitment of individuals and their fate in early-successional fields. In the summer annual populations, there are differences among species in the time of emergence, and these differences are strongly associated with the nature of response of seed germination to temperature (Fig. 5.15). Except in sparse populations, it is expected that early recruits will preempt resources, including space, and suppress later emerging seedlings. Unless the site has unusually strong, early-germinating dominants, such as the giant ragweed (*Ambrosia trifida*), there are no strong indications in early-successional fields that late-germinating species are completely eliminated or are severely disadvantaged. However, late recruits in high density stands suffer some mortality or much lower fecundity than their early-emerging neighbors. Hence, some selection for early germination might occur within populations and among species. However, early germination in the spring in temperate environments is not without cost. Early emergence can mean that the seedlings are likely to experience suboptimally cold nights. If these early germinants are sensitive to cold, they could die or their relative fecundity could be greatly reduced. In such cases, there can be counter-selection for late germination. The two situations would create a strong stabilizing selection and a sharp, narrow window of recruitment. We have found (Drew and Bazzaz 1982) that early-germinating species in early-successional fields, such as *Ambrosia* and *Polygonum* seedlings, can tolerate low night temperature; their stomatal conductance remains equally high after cold or warm nights. Later

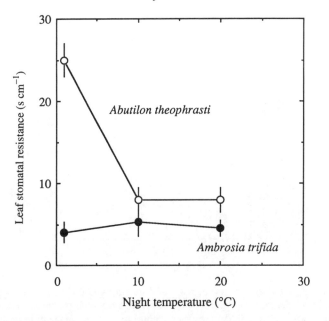

Fig. 5.16. Severe decline in stomatal conductance after a cold night of the late-emerging *Abutilon theophrasti*, and mild decline in the early-emerging *Ambrosia trifida* (modified from Drew and Bazzaz 1982).

germinants in this community, such as *Abutilon* and *Ipomoea*, cannot. Their stomatal conductance sharply declines after cold nights (Fig. 5.16). Recruitment time in the field reflects this sensitivity of seed germination to temperature and sensitivity of seedlings to cold nights. However, because seeds of a given genotype located at different soil depths will emerge at different times, the intensity of selection for both early and late germination can be greatly reduced. Furthermore, we found that in *Polygonum pensylvanicum*, genetic identity was less important than chance in recruitment in high density populations.

Compared to winter annuals, summer annuals seem to suffer less mortality. For example, we observed that *Ambrosia artemisiifolia* population densities remain essentially constant after recruitment. Even in dense populations, they seem to absorb crowding effects through plasticity, and individuals that remain small are able to reproduce. In contrast, *Polygonum pensylvanicum* in the same field and under similar conditions suffers significant mortality. *Setaria faberi* exhibits two peaks of recruitment, and each cohort experiences some death (Fig. 5.17). These patterns of survivorship vary little from year to year. The observation that there is very little

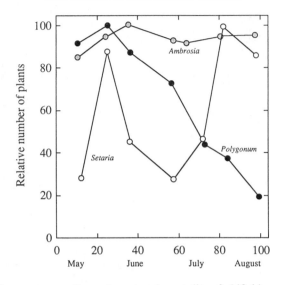

Fig. 5.17. Three patterns of recruitment and mortality of oldfield annuals growing together in the field. *Polygonum* suffers much mortality after the initial recruitment; *Ambrosia* does not; and *Setaria* shows a second peak of recruitment in midsummer (modified from Raynal and Bazzaz 1975*a*).

recruitment of *Ambrosia* and *Polygonum* later in the season may reflect the development of secondary dormancy, which may be a mechanism to replenish the supply of seeds in the seed bank.

In general, there is very heavy mortality during the germination and juvenile phases of a plant's life. Billions of seeds fail to germinate and grow into individuals that reach maturity. Unfavorable environmental factors, pathogens, herbivores, and seed predators eventually kill most of the progeny of an individual (Harper 1977).

Causes of mortality can change dramatically from year to year in the same location, depending on the population dynamics of the agents causing most of the mortality. While any or all causes of mortality can operate in all successional ecosystems, some of these factors can be more prevalent in specific phases of ecosystem recovery. Commonly, trends of the fate of species' populations over successional time have been inferred from annual censuses. In most studies of successional change, emphasis has been placed largely on determining the annual peaks of biomass (or importance) of the species found in different years of succession, and on the rate at which community structure changes (e.g. Pickett 1980). Despite its importance, only a few studies have considered the demography of recruits within a given growing season.

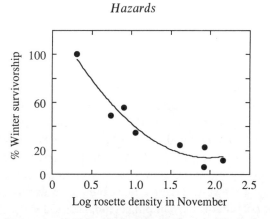

Fig. 5.18. Small rosettes of high density populations in November suffer severe mortality over winter. Size is negatively related to density (modified from Regehr and Bazzaz 1979).

Hazards to recruits in successional habitats

Frost heaving

In temperate regions, the winter annual crucifers and composites can suffer much mortality by uprooting caused by frost heaving over winter. Winter annual species, however, differ in their susceptibility to frost heaving. For example, populations of *Rorippa sessiliflora*, *Erigeron canadensis*, and *Erigeron annuus* of comparable densities and located in the same site suffered respectively, 40, 16, and 6% mortality rate by frost heaving. Examination of root systems suggests that mortality by frost heaving is strongly related to root architecture and extensibility (Regehr and Bazzaz 1979). Well-anchored, large seedlings are more likely to survive over winter. In patches of high-density populations, rosettes grow little in the fall and are therefore poorly anchored during winter. In these patches, very high mortality occurs; mortality has a strong, negative correlation with the size of seedlings at the beginning of the frost season (Fig. 5.18). In this case, frost heaving can regulate population density, select for larger rosettes, and lead to faster growth of the survivors, since they can have access to more resources released from their dying neighbors. Frost heaving of tree seedlings can also be responsible for their death in other exposed sites, such as on blow-down microsites created by uprooting of adult trees (Carlton 1993).

Pathogens and mortality

It is well established that pathogens can cause much mortality in plant populations (Harper 1977, Burdon and Leather 1990), or greatly reduce their fecundity (Alexander 1992). However, the spatial and temporal distribution of pathogens, differences among species in resistance to various pathogens, and the effects of pathogens on the structure of recovering ecosystems and communities is not well known for many systems. Like most other causes of plant mortality, pathogens can vary greatly both spatially and temporally. Variation in disease vectors can be extremely large, and regional and local weather patterns can greatly influence vector behavior and pathogen effectiveness. Disease outbreaks can occur in any system and in some cases can eliminate entire populations over very large geographic regions. For example, in wet habitats (such as the understory of rainforests) damping of fungi can kill many seedlings (e.g. Augspurger 1984).

Pathogenic fungi can be effective even without specific vectors to transport them. Seed-borne fungi seem to be common in early-successional annuals. For example, our work with early-successional annuals (Kirkpatrick and Bazzaz 1979) showed that seeds of some species carry pathogenic fungi within them. We found that the numbers of fungal isolates from seeds of the annuals *Polygonum*, *Datura*, *Ipomoea*, and *Abutilon* were 222, 201, 66, and 13, respectively. Most fungi were specialists with respect to their host plants. The plant species greatly differed in their fungal loads. For example, the percentage of seeds with fungi was 68% for *Polygonum* but only 4% for *Abutilon*. The most prevalent fungal genera on these seeds were *Alternaria*, *Cladosporium*, *Fusarium*, *Stachbotrys*, and *Stemphylium*. These fungi differentially affected seed germination and seedling growth of their hosts. For example, three *Alternaria* isolates completely inhibited the germination of *Abutilon* seed, but had no effect on *Ipomoea* seed germination. Furthermore, different organs of seedlings were differentially sensitive to the same as well as different isolates. *Alternaria* reduced seed germination in *Abutilon* by 66% but completely inhibited radical extension. In *Ipomoea*, the same isolate had no effect on germination but reduced emergence by 70%. Of course, these annuals (like most other plants subjected to pathogens through their evolution) are not without defenses. However, they do differ among themselves in antifungal activities with respect to different fungi. For example, extracts from all tissues of *Ipomoea* had very strong antifungal activates on all 10 fungal isolates tested. Extracts from roots, but not from other tissues, of *Abutilon* had strong antifungal activity, whereas those from *Polygonum* appeared to have only limited activity. We also observed that the fungi influenced seed germination and early seedling development, but apparently did not affect later stages of growth. In these situations, herbivory became a more effective mortality factor on older seedlings and juveniles.

Erigeron, the major winter annual plant in early succession in many temperate regions, can suffer much mortality by the mycoplasma aster yellows. This disease can also substantially reduce seed production of the survivors. In a high infection year, we observed that seed production could be reduced to half of that of a healthy population. Among the winter annuals in oldfields we repeatedly observed that the common crucifer, *Rorippa sessiliflora*, suffers much mortality and reduction of fecundity caused by the fungus *Albugo candida*. We also determined that the degree of infection is density-related: high-density populations suffer proportionally more infection and much higher mortality than low-density populations. Even among the survivors, infection in the high-density populations was six times higher than in low-density populations (Regehr and Bazzaz 1979). Population regulation in this species, like many others, is complex and is likely to be caused by several factors, including pathogens, but soil moisture plays a major role by influencing several factors relevant to mortality. Soil moisture drives recruitment and controls seedling density and size, and can cause death by frost heaving and by the pathogen *Albugo*. Drought, particularly of the upper parts of soil, which also can be localized in the field because of microtopographic relief, can cause the death of entire populations of seedlings.

Pathogens can also devastate populations of some species in mid- and late-successional habitats. The nearly complete elimination of *Ulmus americana* and *Castanea dentata* from the eastern deciduous forest in the United States provides an example of such devastation. Because of difficulties in mastering both population biology and plant pathology, there is very limited understanding of the role of pathogens in regulating recruitment and development of plant populations (but see Burdon 1985, Clay 1990). Differences among locations, years, and species in recruitment, mortality and its causes, fecundity, and other demographic parameters remain fruitful areas for discoveries in population ecology of successional plants.

Seed predators and seedling herbivores

Seed predators seem to be relatively uncommon among the annuals in their first year of succession. However, suppressed individuals of *Ambrosia artemisiifolia*, which occur under the canopy of the winter annual *Erigeron*, can suffer substantial predation in the second year of succession (Raynal and Bazzaz 1975*b*). The release of *Ambrosia* from competition with the winter annuals by their removal in the field resulted in much reduced

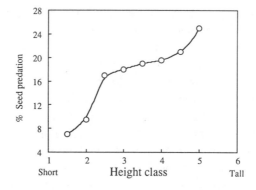

Fig. 5.19. Relation between seed predation and position in the canopy of a stand of *Ambrosia trifida* in the field (modified from Abul-Fatih and Bazzaz 1980).

predation. In *Ambrosia trifida*, we found an inverse relationship between the height of the plant and the level of seed predation. Seeds of taller individuals suffered more predation (Fig. 5.19). In general, seed predation by rodents, ants, etc. may be very prevalent in some forest successions. In fact, the development of a seedling bank through rapid germination of seeds in wet tropical forests may be a means to avoid seed predators (Whitmore 1984, Dirzo 1984, Janzen 1981). In central New York, Gill and Marks (1991) showed that for the common oldfield plants (*Acer rubrum*, *Pinus strobus*, and *Rhamnus catharticus*), seed predation by rodents was the largest cause of mortality. In patchy mid-successional fields, there can be enormous variation in the number of herbivores and seed predators, especially small mammals, as well as in the magnitude of damage to invading tree seedlings.

Herbivory damage to seedlings in early-successional fields may be significant in some years. In one particular year, we observed massive herbivory on seedlings of the annuals in the Illinois field. This severe herbivory coincided with a peak in small mammal populations in the area. In a low herbivory year, Hartgerink (1981) observed that only 5% of *Abutilon theophrasti* seedlings were affected, but up to 95% of the green tissue of herbivorized seedlings was removed. Apparently, because of their relatively high growth rates, early-successional plants can tolerate severe defoliation by herbivores. In a field study, mature individuals that were defoliated as seedlings were smaller than their undefoliated neighbors. However, defoliation accounted for only about 12% of the variance in growth. Under controlled environmental conditions in the glasshouse, defoliated individuals were much shorter than their undefoliated neighbors that experienced competitive release and grew more. In contrast to the field environment, defoliation accounted for more of the variation in plant growth in the less heterogeneous glasshouse environment. Undoubtedly, seedling-scale environmental heterogeneity and compensatory growth played a significant role in reducing the difference between the defoliated and undefoliated groups in the field study.

The timing of tissue loss by herbivory during the life of the individual may also have great influence on the performance of the adults. For example, seedlings of *Abutilon* defoliated when they were one day old and when they were 21 days old produced adults that were greatly different from each other, and from their undefoliated neighbors. Early defoliation resulted in much shorter plants relative to their undefoliated neighbors. Those that were defoliated later were intermediate in height (Fig. 5.20(a)). Especially in fast-growing early-successional plants, defoliation can lead to an enhancement of photosynthesis of the remaining tissue, which can compensate for the carbon gain that would have been lost by defoliation. In *Abutilon*, unit leaf rate (ULR) declined soon after defoliation to 90%, but later increased to 120% of control. With defoliation, there is usually a shift in allocation of biomass toward shoots, which would restore a balance between shoot activities and root activities. There are also changes in canopy display and an increased leaf longevity. The speed of compensation is expected to be higher in early-successional, fast-growing plants than in late-successional, slow-growing species.

The ability to compensate for tissue loss depends on the availability of resources critical to growth, such as light and nutrients, as shown when defoliation occurs in dense stands. Density is a critical factor in differential herbivory in the field and in the outcome of experimental defoliation. In crowded populations especially those with strong asymmetric competition, defoliated individuals become severely disadvantaged with time. For example, in *Abutilon*, severe defoliation (75% of the leaf area removed) had no significant effect on growth and seed production of low-density plants ($16/m^2$), but had great effects on both parameters in high-density plants ($100/m^2$) (Fig. 5.20(b)). Experimental defoliation of *Ambrosia trifida* showed similar trends, in that plants in low densities tolerated a high degree of defoliation without showing much reduction in growth or fecundity, both of which were observed in high density populations.

The relationship of both leaf herbivory and seed predation to succession need more research before generalizations can be made. The origin of the flora, the degree of structural diversity of the vegetation, and its species richness may play important roles in the identity of the herbivores and the

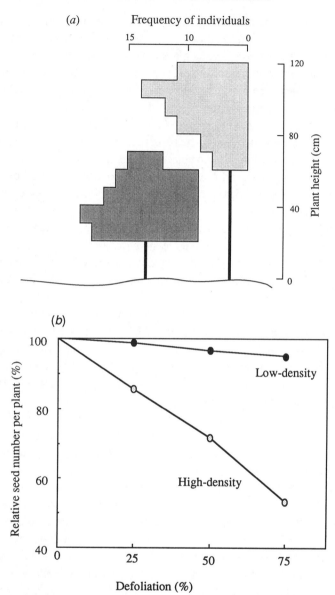

Fig. 5.20. (*a*) Defoliated young seedlings (shaded) in *Abutilon* canopy in the field developed to be much shorter than their undefoliated neighbors (modified from Hartgerink 1981) (*b*) Relative seed number in relation to defoliation prior to flowering in *Abutilon theophrasti* (modified from Lee and Bazzaz 1980).

degree of their impact on the rate and direction of succession. Because of the enormous variation from patch to patch and especially from year to year in the populations of pathogens and herbivores, accurate assessment of their impact requires long-term observation and careful experimentation in the field and under controlled conditions.

An additional cause of seedling mortality in late succession may be physical damage. In tropical forests falling branches seem to be a major cause of seedling mortality (Clark and Clark 1991). The highest risk of seedling mortality appears to be in small gaps rather than in the understory because of falling twigs and branches from dead trees. In many situations in late succession these various causes of mortality do not actually kill individuals but only greatly reduce their growth. Larger individuals of some understory species seem to have developed some resistance to damage caused by falling larger canopy trees. For example, individuals of the understory palm *Astrocaryum mexicanum* made prostrate by fallen trees develop a bend in the trunk and the stem grows upwards (Martínez-Ramos *et al.* 1988*a*). It is not unusual to find negative growth among seedlings and saplings in forest because of branch fall, frost, herbivore consumption, and pathogens. Therefore, the seedling bank of many forests can be made of individuals of comparable height but of very different ages. It is not known whether older individuals will have an advantage over younger individuals of comparable size when resources become available such as by the creation of light gaps above these seedlings.

6

How do plants interact with each other?

Resource acquisition, allocation and deployment, and the equivalency of neighbors

Despite some differences in emphasis among authors seeking strict definitions of the term (see Grace and Tilman 1990), competition is always considered to involve negative interactions mediated through resources in a shared environment. Because interactions among neighbors are not always negative, we suggest the term plant–plant interaction (PPI) as an alternative to describe general relations among neighbors. While competition for pollinators (e.g. Parrish and Bazzaz 1979) and dispersers among successional plants can sometimes be intense, it almost always involves nutrients, water, light, and physical space. Among plants, which normally use the same set of resources, the individual that captures the most resources over time is usually assumed to be the most successful competitor and potentially the most fertile reproducer (Mooney 1976). Therefore, mechanisms of interactions among plants are best understood when the effects of plants on the environmental resources of their surroundings (including themselves and their neighbors), and their response to that pattern of environmental resource availability are known (see Chapter 2). Functionally, competition can be viewed as the modification by neighbors of the processes of *acquisition* of resources, their *allocation* to different parts, activities and functions, and the *deployment* of these parts in space. The ability of an individual to acquire resources depends on (a) the size of resource-acquiring parts, (b) the placement of these parts relative to the resource pools, and (c) the physiological competence of these parts in resource uptake and processing. Species of the same community can differ in their allocation of resources to various structures and activities. For example, growing under the same condition in the early-successional community, *Abutilon theophrasti*

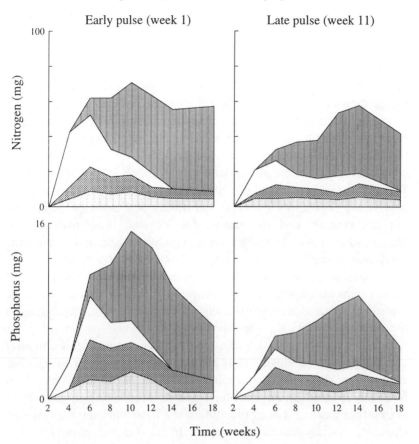

Fig. 6.1. Response of timing of nutrient pulse treatment of the annual *Datura stramonium* showing nitrogen and phosphorus allocation to roots, stems, leaves, and reproduction with early pulse (1 week) and a late pulse (11 weeks) after germination (data from Benner and Bazzaz 1988).

and *Datura stramonium* differ in total biomass, allocation of biomass (especially to leaves), and the amount of nitrogen, phosphorus, potassium, and calcium allocated to fruits and seed (Fig. 6.1). *Datura* allocates relatively more resources to the capture of light (by fast leaf area production and high nitrogen levels) early in growth. In contrast, *Abutilon* allocates more mass, more nitrogen, and more phosphorus to stems (Benner and Bazzaz 1988). Differences in allocation and flexibility in allocation may contribute to co-existence in the early-successional community, as will be shown later. While there has been much work on these three attributes of aboveground parts (i.e. leaf area, leaf display and leaf photosynthetic rates), there has only been limited work on root length, root architecture, and specific root absorption rates. Recently, it has become clear that the placement of plant parts in space and their mode of display (plant architecture) are very important in plant–plant interactions. Recent competition models (e.g. Reynolds *et al*. 1989, Barnes *et al*. 1990, Sorrenson *et al*. 1993) consider measures of canopy architecture as important parameters. In some situations, architecture may be even more critical than physiological processes (such as photosynthetic rates) in determining competitive interactions among neighbors (Küppers 1985, 1989, Caldwell 1987). Plant architecture and several plant physiological traits are intimately interrelated. Branching patterns, leaf angles, leaf mass per area, leaf nitrogen content, and the relationship between these patterns and photosynthetic rates all influence whole plant carbon gain. But despite their great importance to plant function (see Niklas 1988), these architectural and engineering aspects have not often been adequately considered in studies of the mechanisms of plant competition. Gap environments may be exceedingly heterogeneous, especially with regard to light. Depending on their location within the gap, the plants can receive periods of direct radiation at certain times and diffused radiation at other times (Canham *et al*. 1990, Bazzaz and Wayne 1994). Seedling leaf orientation, with regard to canopy openings, is an important determinant of the total amount of radiation received by an individual and its overall carbon gain. High flexibility in seedling canopy orientation and its architecture is a common character of early-successional plants. In the tropical pioneer species *Cecropia obtusifolia, Heliocarpus appendiculatus, Piper auritum,* and *Trema micrantha,* seedlings in gaps differed in leaf number and in leaf size yet their total leaf area per seedlings was similar for seedlings of equal age. A directional correlation test showed that seedling crowns were oriented toward diffused radiation that was received from the canopy gap above, and not toward direct radiation. This was largely determined by non-random orientation of individual leaves (Ackerly and Bazzaz 1995*b*). Due to the lower light saturation ($\sim 1000\,\mu\mathrm{mol\,m^{-2}\,s^{-1}}$) for photosynthesis, relative to direct beam radiation ($\sim 1800\,\mu\mathrm{mol\,m^{-2}\,s^{-1}}$), a leaf may be inclined greatly before incident radiation on it falls below photosynthetic light saturation. In contrast, since diffused radiation is much lower than direct beam radiation, a small inclination in leaves away from diffused light would result in lower carbon gain (Ackerly and Bazzaz 1995*b*). In this case, leaf orientation toward diffused light would be a better carbon gain strategy.

Studies that consider both physiology and architecture have begun to produce a more mechanistic understanding of how species succeed each other or coexist. For example, in hedgerows in Europe, early-, mid-, and late-successional plants differ in the efficiency of their occupation of space (the volume of space gained per unit of dry matter, nitrogen, and phosphorous), with the early-successional plants being least efficient and the late-successional most efficient (Küppers 1985). The replacement of the native *Agropyron spicatum* by the introduced *Agropyron desertorum*, despite similarity in their phenology, photosynthetic capacity, and root weight, is the result of the greater ability of the invader to extract water from the soil because of its higher root density (Caldwell *et al.* 1983). Plants modify the number and size of their parts and their deployment in space in response to neighbors. The modular structure of higher plants allows a great deal of flexibility for efficient placement of parts (see Harper 1985, Franco 1986). Also, by being highly flexible, plants can place their foraging organs in patches that are less contested for by neighbors in terms of resources and space. The plant's degree of flexibility in the deployment of foraging plant parts, relative to the availability of the required resources, is an important component of plant competitive ability. Flexible deployment of parts is a major determinant of habitat choice in plants (see Bazzaz 1991*a*). Such clear habitat partitioning can permit the presence of a wide range of species, and can be a mechanism for coexistence and enhanced species diversity of the community. Because plant communities are usually made up of species with different physiologies and architectures, it is likely that these different species demand different quantities of resources from their environment. It is therefore less likely that neighboring, competing plants are similar in their demand, and that different species will have differing impacts on their neighborhoods. For example, in many situations, the early-successional annual communities include species with differing morphologies and physiologies.

In order to understand the degree of flexibility of allocation and display in the early-successional community, we chose four co-occurring species with contrasting architecture. *Setaria faberi* is a compact bunchgrass with many basal and auxiliary tillers and long, narrow leaves that usually bend back at about the middle of the blade. The stems of *Abutilon theophrasti* are stiff, erect, unbranching, and support large leaves. *Datura stramonium* is erect, with successive dichotomous branches and large leaves. *Polygonum pensylvanicum* is generally branched and has many small leaves (Fig. 6.2). When plants are grown as 'targets' and 'neighbors' (Tremmel and Bazzaz 1993), target biomass, nitrogen content, and seed number are influenced by the identity of neighbors. In targets, leaf birth rates, node birth rates,

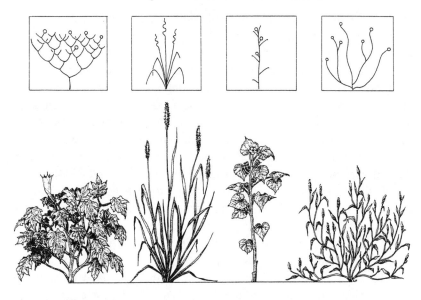

Fig. 6.2. Differences in above ground architecture between the co-occurring early successional annuals (left to right) *Datura*, *Setaria*, *Abutilon*, and *Polygonum*.

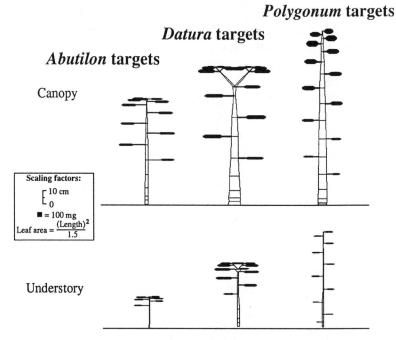

Fig. 6.3. Profile of height and plant mass distribution of *Abutilon*, *Datura*, and *Polygonum* targets grown in similar neighborhoods (modified from Tremmel and Bazzaz 1993).

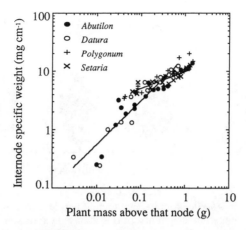

Fig. 6.4. Internode specific weight versus plant mass above the node in *Abutilon* with different competing neighbors (modified from Tremmel and Bazzaz 1995). *Abutilon* allocates differently depending on its neighbor identity.

number of leaves, and mean metamer mass also vary with neighbor identity in the same manner as does total biomass. In turn, development rate, canopy size, and biomass accumulation change with neighborhoods (Fig. 6.3). In these species the relationship between specific petiole weight (SPW) and leaf weight is usually positive. This relationship does not vary with neighbor identity in *Datura*, but varies slightly in *Abutilon*, and varies greatly in *Polygonum*. There are strong positive relationships between internode specific stem weight (SSW) of plant parts above that internode, and this relationship does change in some neighborhoods. For example, in both *Abutilon* and *Polygonum*, SSW changes less with increasing aboveground weight in *Setaria* neighborhoods than it does with *Datura* neighbors (Fig. 6.4). In general, allometric relations between plant parts differ from neighborhood to neighborhood and among plants of different sizes of the same species in the same neighborhood. For *Abutilon*, the allocation of biomass at the upper nodes is severely reduced in the smaller individuals and leaf shape changes as well. In contrast, the shape of leaves and number in *Polygonum* canopies are similar across all size classes, but plant height and node number decrease with weight (Fig. 6.4). Plants growing with *Polygonum* or *Setaria* as their neighbors have heavier plant parts and denser support structures. Leaf angles, leaf area, and specific leaf area are also extremely variable in different neighborhoods. Also, different weight classes of *Abutilon* differ among themselves in response to neighbors. Suppressed *Abutilon* individuals have a large reduction in biomass allocation

How do plants interact with each other?

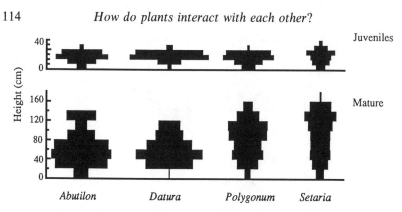

Fig. 6.5. Changes in *Abutilon* architecture (height and leaf area) in response to the identity of the competing neighbors during a growing season (modified from Tremmel and Bazzaz 1993).

to stem, but only a small reduction in node birth rate. Thus, they are quite short and have near monolayer canopies (Fig. 6.5).

In the early-successional community, species can occur in patches of different densities and identity of neighbors. In *Abutilon*, change in petiole length and a reduction of blade size are pronounced responses to neighborhood density (McConnaughay and Bazzaz 1992*b*). Leaves generally exhibit more variability than either petioles or internodes, the latter being the least responsive to neighbor identity. In all of our studies, conspecific neighbors have significantly greater effects on growth than do heterospecific neighbors. All the species show great flexibility in aboveground allocational and architectural traits, resulting in different plant shapes. In some cases, in multispecies interactions, there is also notable decoupling between vegetative growth and sexual reproduction. It is known, for example, that suppressed individuals of *Abutilon* abort most of their flowers, especially under a *Polygonum* canopy (Pickett and Bazzaz 1976). Except for the persistence of seed banks in soils, which play a major role in maintaining the species in a site, this decoupling can have significant impact on community composition and its phenotypic diversity following future disturbance (which will be explained later).

By having different architecture, the species can differentially modify resource availability for their neighbors. For example, light penetration expressed by the extinction coefficient, differs in these canopies, with *Setaria* allowing the most and *Polygonum* allowing the least (Fig. 6.6). In contrast to other annuals, *Datura* has a more fixed architectural design, with rigid, dichotomously branching stems. Thus, the identity of *Datura*'s

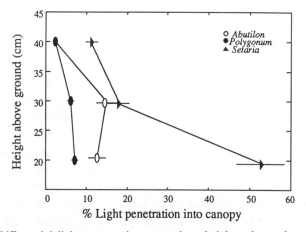

Fig. 6.6. Differential light penetration to various heights above the ground in *Abutilon*, *Polygonum*, and *Setaria* stands (modified from Tremmel and Bazzaz 1995).

neighbors has little influence on its architecture. Because of this rigidity and the fact that flowers are born at the points of branching, even suppressed individuals reproduce successfully as long as they attain branching size. In contrast, *Polygonum* is the most variable of the species in allocational and architectural traits. In *Polygonum*, crowding causes some of the stems to meander among its neighbors and escape to locations of less plant biomass. This flexibility allows seedlings of this species to escape from their neighbors and forage in areas of less contested resources such as small canopy gaps with high light levels (McConnaughay and Bazzaz 1992*b*). Furthermore, in dense stands suppressed individuals of *Polygonum* do not usually die, perhaps because of their relatively low light compensation point of photosynthesis (Wieland and Bazzaz 1975). Unlike *Abutilon*, even small individuals of this species are able to produce seed.

Tissue nitrogen content is unaffected by neighbor density, and except with *Polygonum*, target biomass was poorly correlated with neighbor leaf area. However, for all species, target biomass correlates highly with the proportion of potential light interception blocked by neighbor canopies. Thus, it is the amount of reduction in light availability that influenced performance, irrespective of the identity of the neighbor. In other words, competition among neighbors occurs mainly through modification of resource availability.

Plants influence their neighbors by occupying space *per se* as well as by depletion of resources. Because space in and of itself can also be contested by neighbors, physical restriction of allocation of biomass and its deployment, such as might happen in crowded stands, can reveal much about architectural flexibility and its consequences for growth and reproduction. For example, experimentally decreasing available physical space for plants by inserting inert glass rods in developing stands of the early-successional annuals, generally results in smaller plants with fewer, smaller leaves and generally high leaf area ratios (LAR). The annuals differ in the degree of their sensitivity and in allocation of biomass to simulated neighbor 'shoot' density (Fig. 6.7(a)). For example, proportionally more biomass is allocated to leaves in *Chenopodium*, to stems in *Polygonum*, and to petioles in *Abutilon* (McConnaughay and Bazzaz 1992*b*). Interactions among neighbors in crowded situations depress growth of all species because of differences in allocation. Under controlled conditions, with specified soil resources, competition between the annuals *Amaranthus retroflexus* and *Abutilon theophrasti* causes a decline in growth of both species, but by different mechanisms. In *Abutilon*, leaf area ratio (LAR) decreases, whereas in *Amaranthus*, unit leaf rate (ULR) decreases. In *Amaranthus*, root/shoot ratio decreases but there is a much higher rate of nitrogen uptake per unit of root. *Abutilon* is relatively more successful than *Amaranthus*, mostly due to its large starting capital, i.e. its large seed (Bazzaz *et al.* 1989).

It is now becoming clear that belowground allocation, deployment, and competition in nutrient and water uptake are also critical in plant–plant interactions. It has been known for a long time that plants, even within the same habitat, exhibit very different belowground root placement (Weaver 1919, Wieland and Bazzaz 1975, Parrish and Bazzaz 1976, Caldwell *et al.* 1987), different above- and belowground architectures (e.g. Caldwell 1987, Barnes *et al.* 1990), and different efficiencies in nutrient and water uptake rates (Marschner 1986). Mechanical constraints also influence the allocation of mass and allometric relations between plant parts. Because of technical difficulties in studying the placement of roots and rhizomes in the soil, belowground architecture and its consequences to competition and community structure still receive only limited attention (see Fitter 1987, Berntson 1994). There is no reason to suspect that underground architecture is less relevant to plants than is aboveground architecture. Both parts of the plant forage for resources necessary for growth and reproduction. As is the case for aboveground parts, plants tend to place their underground foraging organs in areas of less contested resources and adjust their root architecture to reduce interference. Roots are also known to avoid each other when they are in close proximity (e.g. Mahall and Callaway 1991). The flexibility in deployment of belowground parts may also act to maximize soil resource capture in competitive environments. Additionally,

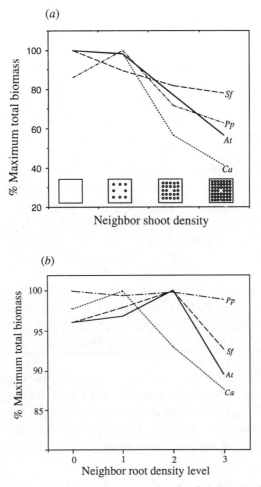

Fig. 6.7. (*a*) Response of biomass of the annuals *Abutilon, Setaria, Polygonum,* and *Chenopodium* to increasing preemption of aboveground space by inert objects. (*b*) Response of biomass of the early-successional annuals to the preemption of underground space by fake roots made of plastic coated wires (modified from McConnaughay and Bazzaz 1992*a,b*). *Pp, Polygonum pensylvanicum; Sf, Setaria faberi; At, Abutilon theophrasti; Ca, Chenopodium album.*

more occupation of space may modify root allocation of neighboring plants and in turn affect shoot activity and whole-plant performance. Using simulated roots made of plastic-covered electric wire, McConnaughay and Bazzaz (1992*b*) have shown that species of the early-successional community differ greatly in their response to the degree of physical occupation of both belowground space and resources (Fig. 6.7(*b*)). For example, when given larger rooting space, *Abutilon* increased allocation to reproduction relative to vegetative tissues, and produced significantly larger seeds. In contrast, *Setaria* responded to changes in belowground space mainly by phenological shifts. It flowered early and had greater reproduction in restricted belowground space (McConnaughay and Bazzaz 1992*a*). Fragmenting underground space, as might occur when competing plants are present close to each other, can result in altered root architecture, including branching pattern, branch root length, and branching angles. Thus, physical space, irrespective of the amount of available resources, influences the performance of these colonizing annuals both above- and belowground.

Specific absorption rate (SAR) can change with root type, position and age, perhaps as much as leaves differ in their photosynthetic rates with age and position. Thus, expressing resource gathering ability on the basis of total root weight or total area can sometimes be less informative for understanding the mechanisms of whole plant underground resource capture than considering length, area and the efficiency of capture of the resources in question (e.g. Berntson *et al.* 1995). Although there are clear differences among them, early-successional, fast growing species seem in general to be architecturally very flexible, both above- and belowground. There is little doubt that this flexibility is an important component of their overall life history strategy.

The above discussion concerned coexisting species that occur in the same habitat but do not usually replace each other in succession. While their interactions are not directly implicated in successional change, they can indirectly influence the patterns of species distribution of plants that succeed them (as will be shown later). Most of these annuals are eliminated quickly from successional fields as succession continues, but some individuals can also occur in small gaps created largely by mammals (McConnaughay and Bazzaz 1987, 1990). Some of these individuals can be large and contribute to the seed bank. In such gaps, with their severe restrictions of resources and physical space, both the identity of the competitors and the levels of resources in the gap can influence the outcome of plant–plant interactions and the relative contribution to the seed bank of competing species (Fig. 6.8). In field experiments with several of these annuals in gaps, under resource-limiting conditions, we found that conspecific neighbors reduce plant growth more than do heterospecific neighbors. There was also no consistent competitive hierarchy of species in the system in these small gaps. Furthermore, although the identity of neighbors influenced plant survivorship and growth differently, it had only limited effect on seed production; perhaps this is another expression of flexibility. Apparently,

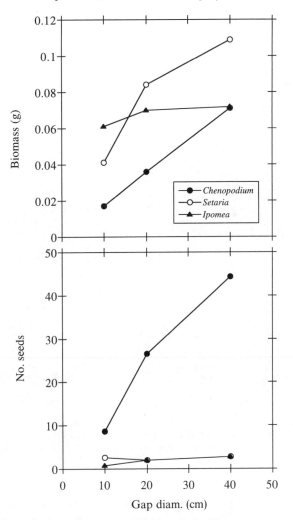

Fig. 6.8. Differential response of the early-successional annuals to gaps of different sizes created in a mid-successional community dominated by the grass *Poa pratensis* (modified from McConnaughay and Bazzaz 1987).

under these limiting situations in gaps in older vegetation, the nature of plant–plant interaction among the early-successional annuals do differ from the interactions in younger fields.

Difference in acquisition, allocation, and deployment of resources, among species that do replace each other during succession, do exist and may contribute to replacement itself. In mid-successional fields, *Aster* and *Solidago* can occur together in patches, sometimes on a background of

Aster lanceolatus   *Solidago canadensis*

Fig. 6.9. Difference in architecture (rhizome length, branching patterns, leaf persistence) between *Aster* and *Solidago* (modified from Schmid and Bazzaz 1994).

grass species. They are both important in successional fields. Commonly, *Solidago* invades and ultimately replaces *Aster*. Although the two species are the same overall size, these clonal species differ appreciably in their architecture (Fig. 6.9). *Aster* has branched shoots, long rhizomes, and a spreading genet architecture, whereas *Solidago* has unbranched monopodial ramets, short rhizomes, and compact genets. These basic differences in architecture have demographic, and physiological implications that are important in the way these species compete and in their successional replacement (Schmid and Bazzaz 1990, 1994). The more open canopies of *Aster* genets can be invaded by the compact genets of *Solidago*, but the reverse is less common. Thus, competition for aboveground resources in *Aster* occurs among branches and leaves of the same genet, of different genets, or of different species, whereas competition among leaves in

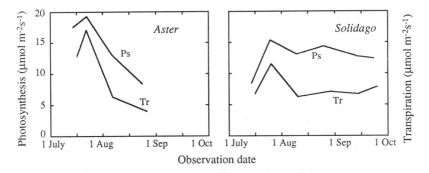

Fig. 6.10. Changes in photosynthetic (Ps) and transpiration (Tr) rates with age of leaves in *Aster* and *Solidago* in the field (modified from Schmid and Bazzaz 1994).

Solidago genets is mostly among themselves. Because of much branching, *Aster* has an exponentially increasing leaf population. This leads to a specific architecture: the presence of many leaves at the top of the canopy, a high leaf turnover rate, and a shallow monolayer canopy. In contrast, *Solidago*, with an unbranched shoot, grows rapidly in height, produces a linearly increasing leaf population, and exhibits a slow leaf turnover rate, leading to the formation of a deep canopy. Leaves of *Aster* are cheaply built, with low mechanical strength and low nitrogen concentration. They reach their peak photosynthetic rate early, which then quickly declines with no significant translocation of NPK before they drop (Fig. 6.10). In contrast, leaves of *Solidago* are built to last. Their photosynthesis rate decreases only slightly with age. Relative to upper leaves, lower leaves of *Aster* experience a greatly reduced rate of photosynthesis. Lower *Solidago* leaves are more efficient in low light relative to lower *Aster* leaves (Fig. 6.11). Because of the differences in the structure of the canopy, leading to differing degrees of potential penetration by other plants, leaf demography in *Aster* was more strongly influenced by competing grass species in experimental plots than it was in *Solidago*. In these two species, as in many others, plant architecture, leaf demography, carbon gain, and allocation patterns are intimately related to each other and are critical to the way plants interact and replace each other in succession.

What do these interactions mean for succession? Traits of neighboring individuals either mesh well (cooperate), resulting in continued co-occurrence of individuals, or compete strongly, resulting ultimately in replacement of individuals. The notion of combining ability (*sensu* Aarssen 1983) of commonly co-occurring species is an expression of this cooperation. Exploring resource uptake, allocation and deployment and their degree of

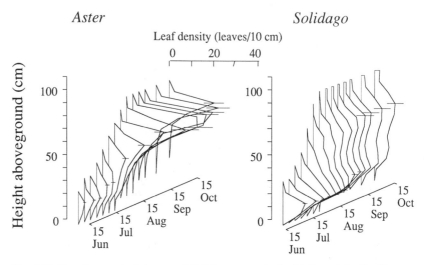

Fig. 6.11. Development of *Aster* and *Solidago* canopies by birth and death of leaves over a growing season. Lower leaves are dropped faster in *Aster* than in *Solidago* (modified from Schmid and Bazzaz 1994).

flexibility would allow a better understanding of how different plants compete with each other, rather than merely 'black-boxing' the mechanism of interactions among them and considering their final outcome. These two different approaches have different aims: the first is necessary for a deeper understanding of how plants interact with each other, and the second is sufficient for constructing certain models of plant–plant interactions to predict their outcome.

Early-successional habitats usually have well developed and rather uniform canopies formed by the rapid growth of the annuals that dominate them. In these plants, different architectural designs are possible because of a high degree of flexibility. As succession proceeds, the canopy becomes more patchy as invading shrubs and trees begin to emerge above the herbaceous canopy. Because of the process of gap creation and filling, mature phase forests also have less uniform canopies, with complex architecture caused largely by the presence of several life forms and high species diversity. Some canopy models assume that communities are designed so that the growth of the entire stand is maximized (Russell *et al.* 1989). The plants have to 'cooperate' more than 'compete.' However, a genotype or species that is more fecund can potentially replace the others in that community. From an evolutionary perspective, individual productivity and fecundity are optimized in natural communities, and neighbors compete for limited resources in a struggle to increase their own relative fitness. Long-term ecosystem level productivity must be viewed as the outcome of the evolution of species populations to maximize their own life-long fitness rather than the maximizing of overall productivity. Overall ecosystem productivity results from the interaction of many species with very different traits and varying degrees of architectural flexibility that, in plants, is greatly aided by their modular construction. Understanding the mechanisms of how species replace each other in succession requires knowledge of the relation of form to function in the major species.

The identity and equivalency of neighbors

Plant species diversity generally increases with succession and may often reach a maximum before the attainment of a stable community with low levels of disturbance (Chapter 3). While we can assume that the identity of neighbors is more predictable in low diversity than in high diversity ecosystems, dispersal patterns, environmental patchiness, and localized seed bank dynamics complicate this predictability and make it highly scale-dependent. Moreover, as species diversity increases during ecosystem recovery, the balance between intraspecific and interspecific competition can shift, with the former more common than the latter in early-successional communities. However, irrespective of the age of successional fields, competitive neighborhoods can differ greatly in their complexity. Neighbors can be influenced by both the density and the identity of the interacting species of competitors, and possibly by the number of these species (Bazzaz and Garbutt 1988). For example, does the presence of a few or many species, irrespective of density or total biomass, influence competition? Competition among neighboring plants can be symmetrical, size-symmetrical, or asymmetrical. In symmetrical interactions, neighboring plants have an equal influence on each other's growth and reproduction, regardless of differences in architecture, or physiology. This is probably the least common type of interaction in nature. More commonly, interactions among neighbors are size dependent, as individuals are generally assumed to accumulate resources in proportion to their biomass (Weiner 1985, Gaudet and Keddy 1988). In asymmetric interactions, larger individuals acquire a disproportionately larger amount of a resource relative to their size (Thomas and Weiner 1989*a,b*). The analyses of symmetry among plants have been largely concentrated in single species stands, mostly of annuals. Despite the importance of size, the development of size hierarchies, and the nature of plant–plant interactions within these hierarchies, has not been adequately addressed with regard to succession.

However, some ecologists (e.g. Aarssen 1983, Ågren and Fagerström 1984, Shmida and Ellner 1984, Hubbell and Foster 1986) have suggested that competitive effects are more or less equivalent for species in many plant communities. Because they have broad responses and a high degree of response overlap on critical environmental gradients, early-successional plants can be considered more or less equivalent in their interactions (see Bazzaz 1987). Parrish and Bazzaz (1982a) and Goldberg and Werner (1983) investigated the equivalency of competing successional plant species that have the same growth form, and argued that competitive interactions are not usually species-specific. When equivalency is high, different competitors are expected to have a similar per unit biomass effect on their neighbors. Size becomes the major determinant of the interaction; i.e. competition is size-symmetrical. In an early-successional community, Miller and Werner (1987) found a negative correlation between competitive effects and responses. They found that *Ambrosia artemisiifolia* has the largest effect on the other species in this system and the smallest response to their presence. The mean effect on their neighbors of *Ambrosia*, *Agropyron*, *Plantago*, *Lipidium*, and *Chenopodium* decreased in that order. Thus, the interaction among these species was highly asymmetric. There was also a lack of specificity in these interactions. Except for *Chenopodium*, all species appeared to have the same per unit biomass effects on the focal species. Goldberg (1987) investigated competition between *Solidago canadensis* and seven co-occurring herbaceous plants in an oldfield in Michigan. She showed that the average growth of *Solidago* was reduced by 17–62%, depending on the identity of its competitors. However, there were no consistent patterns in competitive effects. Generally, the competitor neighbor identity was unimportant. Also, per-individual effects on *Solidago* target biomass were strongest for conspecific competition and for species with the same growth rate as *Solidago*. As shown earlier in this chapter, neighbors in Illinois oldfields are not equivalent in their impact on one another; however, this study shows that, relative to plants from later-successional communities, early-successional neighbors are more equivalent.

Experimental tests with early-successional communities have confirmed this. In a diallel competition experiment, Parrish and Bazzaz (1982a) found that the mean biomass of an individual competing with another individual of the same species was not different between the early-successional annuals and the late-successional herbaceous species. In the late-successional species, however, the mean biomass of an individual was significantly higher in heterospecific pairs. Furthermore, the ratio representing the coefficient of variation of mean weight of an individual of a species in a

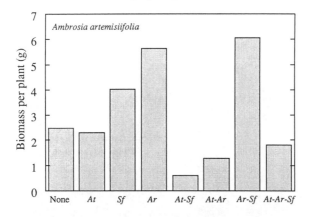

Fig. 6.12. The response of *Ambrosia artemisiifolia* to the identity of neighbors and their diversity under controlled environmental conditions (modified from Bazzaz 1990a). *At, Abutilon theophrasti*; *Sf, Setaria faberi*; *Ar, Amaranthus retroflexus*.

mixed stand over the mean in pure stands was much lower in the early-successional community than in the late-successional community. That is, there were smaller differences in performance among species of the early-successional community. Also, *Amaranthus retroflexus* and *Setaria faberi*, C_4 plants in the early-successional community, responded in competitive situations to any combination containing C_3 plants in the same way (Bazzaz and Garbutt 1988). However, both the identity and the diversity of neighborhoods influenced target performance in some species more than others. *Ambrosia artemisiifolia* biomass is reduced greatly in high diversity neighborhood whereas *Abutilon* biomass is not (Fig. 6.12).

In the mid-successional community where *Solidago* and *Aster* compete with other co-occurring species, especially grasses, we also found that a similar amount of neighborhood biomass, when produced by different neighbor species, had different effects on the two target species (Schmid and Bazzaz 1992). Thus, within groups of similar-sized species, architecture can account for variation in competitive ability. *Poa* neighborhoods, which have high root density, caused dense aggregation of rhizomes in both *Aster* and *Solidago*, whereas both species had more spreading rhizomes in neighborhoods with neighbors of low root density, such as *Dactylis*. Resistance to invasion by *Solidago* will therefore be stronger in *Poa* patches than in *Dactylis* patches. It appears that in this mid-successional community, neighbors are less equivalent than those in the early-successional community and more equivalent than species of the late-successional community.

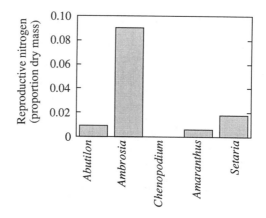

Fig. 6.13. Differences in seed nitrogen content of early-successional annuals growing in similar soil conditions.

Therefore, neighbor equivalency appears to decline during ecosystem recovery.

There are several reasons why we should not expect a high degree of equivalency in any plant community. While plant species within a community could be similar in their response on one or more resource gradients, they are unlikely to be similar on many gradients (see Bazzaz 1987), and therefore must experience varying strengths of interactions when growing in proximity to each other. Species differ in their allocation and architecture, thus allowing different light penetration and unoccupied above- and belowground space. Species can differ greatly in the concentration of mineral elements; therefore, one gram of a species with low nutrient content is unlikely to have the same influence as one gram of a competitor with a higher concentration of that element (Fig. 6.13). Finally, species differ from each other in the efficiency with which they use these elements. For example, under the same conditions, relative growth rate (RGR) is higher in *Abutilon* than in *Amaranthus* at the same nitrogen concentrations (Coleman and Bazzaz 1992).

Despite the broad responses of many early-successional plants, their relative equivalency as competitors, and the great differences in competitive outcome depending on the environment, there are some examples of superior competitors among them. Among the common annuals, *Ambrosia artemisiifolia* is a particularly strong competitor as mentioned earlier. In controlled experiments, *Ambrosia* reduced the height growth of *Aster* by three times more than did any of the other co-occurring annuals in the experiment (Peterson and Bazzaz 1978). In another experiment, *Ambrosia* seed accumulated about five times more nitrogen relative to its other co-occurring annuals (Garbutt *et al.* 1990). *Ambrosia artemisiifolia* possesses a variety of attributes that make it a very successful plant. Its germination behavior (that is keyed to disturbance), its high carbon gain capacities, its broad response to elevated CO_2 and temperature, and its resistance to water stress have been implicated in its superiority over other summer annuals in oldfields (Bazzaz 1974). A congener, the giant ragweed *Ambrosia trifida*, dominates the community and decisively out-competes all other plants when present. Extensive life history and physiological studies have shown that within its guild, this species has the largest seeds, the earliest emergence, the highest tolerance to cold nights in the early spring, the largest cotyledons, the fastest growth rates, and the highest leaf area index (Bazzaz 1984*b*). It also fixes a significant amount of carbon in its photosynthetic flowers and seeds, greatly lessening reproductive cost, in terms of carbon, to the remainder of the individual (Bazzaz *et al.* 1979). Removal experiments in the field have shown that the presence of *Ambrosia trifida* lowers community diversity by 90% but increases productivity eight-fold (Abul-Fatih and Bazzaz 1979*a*). Except for limited seed dispersal abilities, the species would overwhelmingly dominate oldfield communities in most of the eastern United States and perhaps other disturbed habitats in other temperate regions. For example, the species is currently found in some locations in Japan (personal observation). With increased disturbance, and inadvertent seed dispersal, this species can become important in temperate environments worldwide, causing problems to hay fever sufferers.

In the winter annual community of the southeastern United States, *Heterotheca latifolia* is rapidly replacing *Erigeron canadensis* as the major pioneer species in oldfield succession. Tremmel and Peterson (1983) have shown that competition between the two species in the field is so one-sided and intense that when seeds of the two species were sown in equal numbers and high density, all seedlings of *Erigeron* died and a pure stand of winter rosettes of *Heterotheca* developed. Exotic species, with special attributes that make them particularly superior competitors relative to the resident species, such as *Myrica faya*, which fixes nitrogen, can easily invade successional habitats (Vitousek and Walker 1989) and change the nutrient dynamics of the system and the rate of succession. The entry of such species into an ecosystem may have a great impact on successional patterns.

7

Plant–plant interactions and successional change

Ecologists have been speculating for decades about the mechanisms and causes of succession (see McIntosh 1980, Miles 1987, Pickett *et al.* 1987). In most cases competition was assumed to be a major factor in species replacement. Competitive superiority is generally taken to mean the relative ability of a plant to take up resources so that an individual deprives its neighbor (competitor) from obtaining these resources and therefore reduces its growth and fecundity (Chapter 6). As early as 1950, C. Keever carried out detailed experiments to evaluate the causes of replacement of the annuals *Erigeron canadensis* and *Ambrosia artemisiifolia* by the perennial *Aster pilosus*, which was in turn replaced by the grass *Andropogon virginicus* in oldfield succession. Keever (1950) concluded that for these herbaceous plants, differences in dispersal patterns, germination time, growth rate, drought resistance, light requirements, and autotoxicity determine the sequence of species in succession. Competitive superiority in the usual sense is important *per se*, but it is not the only factor in this system, nor perhaps in most successional ecosystems.

If competitive superiority is viewed as the ultimate success of a species relative to other species that occupy the same location, then the whole process of succession will simply be the replacement of species by the next, relatively more superior competitor. In this view, then, competition is the force that drives succession. But, if we believe that species mature, reproduce, and disperse individualistically and differentially in space and that they have different abilities to take up and use resources, then successional sequences will be largely determined by propagule arrival time and the time each species requires to grow, reproduce, and disperse. Plant–plant interactions, in this case, modify these intrinsic activities of neighbors but do not completely govern them (Bazzaz 1990a). Therefore, the success of species in invading a habitat and replacing earlier occupants depends on a suite of life history traits that includes the species relative competitive competence in resource acquisition, allocation and deployment.

Competition among neighbors for limited resources, such as light, water, nutrients, and space, has been among the most widely studied subjects in plant ecology (see Grace and Tilman 1990). The role of competition in plant communities has always been assumed to be central (Lomnicki 1988, Connell 1990), and has been considered to operate largely through resource depletion (Werner 1979). However, some researchers (see Rice 1984, Williamson 1990) have recognized the possible role of chemical inhibition as a factor in species interactions and replacement and the resulting successional change. Many of these chemicals appear to interfere with the availability of nitrogen in soils and its utilization by the plants (Rice 1984). Plant–plant interactions can also be greatly influenced by other biotic interactions, such as herbivory (Connell 1990). Experimental tests of the role of competition in recovering ecosystems have been especially informative when community changes were assessed after the removal of species from the community.

Interactions among neighbors are not always competitive; they may be facilitative or neutral. The corresponding outcomes – inhibition, facilitation, and tolerance – have all been considered important driving forces in successional change (Connell and Slatyer 1977). However, interactions among neighboring individuals are rarely of a single type. In fact, negative and positive interactions among individuals may occur simultaneously. Inferior competitors for above- and belowground resources or late recruits can become suppressed through competition for resources as the season progresses, but they may, in the meantime, become dependent on the winners for survival. For example, if the smaller individuals develop a shade-adapted physiology (e.g. thin leaves, etiolated stems, low *Rubisco* concentration and activity, high chlorophyll content, and higher photosynthetic efficiency at low light, etc.; Givnish 1988), the death of the taller dominants may expose the smaller individuals to stressful and possibly damaging environments such as high light, high temperature, low humidity, and high wind, for which they are not adapted. Suppressed individuals in many early-successional communities may also depend on their bigger neighbors for physical support, and, in some cases, this dependency may allow the smaller individuals to allocate relatively more mass to seed production (Abul-Fatih and Bazzaz 1979*b*). Plants in a variety of community types are known to support their neighbors (Holbrook and Putz 1989, Thomas and Weiner 1989*a*). Many understory herbs in the forest compete with the trees for water and nutrients, but depend on the trees for shade. In some forests, when the overstory is removed, the forest can decline as the vegetation becomes exposed, dying back under high light exposure. Neighbors compete, lean on each other, protect each other, attract pollinators in aggregation, and share mycorrhizae. They interact in various ways and not only by competition. Therefore as we mentioned in Chapter 6, the expression 'plant–plant interaction' (PPI) is more appropriate than 'competition' for describing relationships among neighboring plants.

Plants interact when individuals are in such close proximity that each may influence the performance of the other in terms of altered growth, survival, and reproduction. Competitive interactions among plants involve the differential *acquisition* and *allocation* of carbon, nitrogen, and other minerals, and the *deployment* of plant parts in space (plant architecture) to optimize acquisition both above and below the ground. Changes in physiology, allocation patterns, and architecture caused by negative interactions often render the target individuals less fecund than they would be in the absence of neighbors. Neighboring individuals can change each others' environments, usually reciprocally, mostly asymmetrically. Species of different life forms and different successional positions are expected to differ in these characteristics.

Because the plant is a balanced system (Mooney 1972, Bloom *et al.* 1985), it adjusts its allocation to its resource-gathering organs such that each supplies sufficient resources, so that the plant as a whole can grow and reproduce. Soil resources, such as water and nutrients, must match aboveground resources, such as light and carbon dioxide. The individual plant allocates manufactured resource products (e.g. carbohydrates, proteins, fats, etc.) to these various compartments according to need, in order to keep the system in balance. Water uptake from soil is scaled to water loss from leaves. Thus, there must be some allometric relation between leaf activity and root activity. Moreover, nitrogen uptake by roots is scaled to nitrogen levels in the shoot, which in turn is related to the photosynthetic rate of the shoot, which is related to the transpiration rates of the shoot, and so on. While biomass has been used to quantify ratios of plant parts, it does not directly indicate activity if roots and leaves differ among themselves in their uptake efficiency and/or photosynthetic rate. It is therefore the *activities* rather than the *weights* that are to be balanced. For example, PN-use efficiency, i.e. unit carbon gain per unit of nitrogen in the leaf (especially in *Rubisco*, PEP Carboxylase in C_4 plants, and chlorophyll) can vary greatly among plants and in different environments. Also, root specific absorption rate (RSR) can also vary among species. The only way for a plant to be a strong competitor aboveground and poor competitor belowground is to be able to do much more growth aboveground with much less of the underground resources relative to its neighbors (e.g. have a much higher nitrogen-use efficiency) and also be able to photosynthesize maximally even at more negative leaf water potential. This is an unlikely situation in plants. However, largely under the influence of the changing local environment, shoot and root ratio can occasionally fall out of balance. For example, high transpiration rates during the morning hours can cause a reduction in leaf water potential because of the inability of roots to supply sufficient water from the soil. A balance can be partially achieved in the short term by stomatal closure, causing what is referred to as a midday drop in transpiration and photosynthesis, commonly observed in species of sunny habitats. Increased biomass allocation to roots and a change in their architecture to forage more broadly for water may restore this balance in the long term. Also, some plants shed their lower leaves to reduce transpiring surface under water-limiting conditions. In very fast growing early-successional annuals such as *Ambrosia trifida*, allocation of biomass to shoots and roots continues to shift. Over the growing season the plant allocates more to one structure early in life and to the other later.

Competition and change: how species replace each other during succession

Shortly following the invasion of an open habitat, plants begin to interact with each other. In fact, interactions among neighboring plants may actually start before individuals begin to share the same resource pools or are in actual contact with each other. Plants alter the wavelength of lights reflected even to remote neighbors (Ballaré *et al.* 1987, Novoplansky *et al.* 1990). Despite some mechanisms for avoidance, the intensity of these interactions increases as the vegetation develops and biomass increases. These interactions together with life history traits usually result eventually in the local extinction of the target species and the dominance of what is assumed to be a superior competitor. Complete competitive exclusion may ensue over a large area, although this is rarely observed in nature on any meaningful scale. Most early-successional species, however, are found in very low number in late-successional communities.

During ecosystem recovery, plant neighborhoods and the interactions they encompass change over time. In forest regeneration, some individuals in the understory may experience alternating periods of low and high competition from neighbors, especially for light, as gaps above them open and then close. These individuals attain much of their growth during the former periods and tolerate shortages during the latter until they finally reach the canopy. In these species some degree of both opportunism and tolerance is needed to finally ascend into the canopy and attain reproductive maturity.

Interaction in monospecific patches in early-successional fields

Patches of single species populations can be found in most stages in ecosystem recovery. However, interference among conspecifies has not been directly implicated in successional change, except in cases where autotoxicity is assumed to result in eventual local extinction, as in the case of *Erigeron canadensis* and its replacement by *Aster*. In successional habitats, an individual usually interacts simultaneously with its siblings, with other members of the same species. It can also interact with species of different taxonomic affinities and growth habits. Except for extreme clumping of individuals caused by dispersal patterns, the probability that an individual interacts with individuals of different species increases with the increase in species diversity that occurs as succession proceeds. Nevertheless, intraspecific competition may be a significant selective force in evolution, especially for seed dispersal, timing of seed germination, early growth, and accelerated reproduction in early-successional species. In these stands, size hierarchies develop. With time, suppressed individuals usually suffer most of the mortality and dominant individuals usually contribute most of the seeds to the next generation. If dominance and suppression are non-random events, then the resulting life history traits may be important to species replacement in succession.

We have studied the nature of intraspecific (conspecific) competition in several major species in oldfield succession: the winter annuals *Rorippa* and *Erigeron*, the summer annuals *Ambrosia* and *Polygonum*, and the clonal *Solidago*. In *Erigeron*, we found that conspecific competition significantly regulates population density mainly by influencing the size of rosettes before they enter winter. In high density stands (average rosette density of $320\,\mathrm{m}^{-2}$), each seedling had limited access to resources and therefore grew little and formed relatively small rosettes. Mean rosette size in November at this site was significantly less than that in medium ($120\,\mathrm{m}^{-2}$) and in low density sites ($10\,\mathrm{m}^{-2}$) in the same field. Within-and between-site mortality over winter was negatively correlated with rosette size (see Fig. 5.18). Smaller rosettes had a higher probability of death than did larger rosettes, as many of these small, less well-anchored rosettes were eliminated by frost heaving during the winter. Competition among young rosettes did not lead directly to much mortality. It did contribute indirectly, as plant interactions

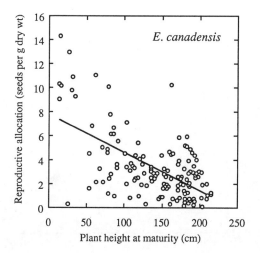

Fig. 7.1. Relationship between plant height and reproductive allocation in the early-successional winter annual *Erigeron canadensis* (modified from Regehr and Bazzaz 1979).

controlled the amount of growth attained by individual rosettes in autumn. Furthermore, rosette size determined the contribution of the individual to future generations: bigger individuals, derived from larger rosettes, had much higher seed production than their shorter neighbors, derived from smaller rosettes (Fig. 7.1). Populations of other major winter annuals, e.g. the composite *Erigeron annuus*, behave similarly despite some differences in phenology (see Bazzaz 1984*b*).

The common crucifer in early-successional fields, *Rorippa sessiliflora*, emerges in the summer before the composites, usually after heavy rains. The species forms many small monospecific patches consisting of a few to several hundred individuals. Like the composites, its seedlings grow in the fall and overwinter as rosettes. Small rosettes suffer much mortality over winter. The survivors bolt, flower, and set seed early in the spring, before the composites have begun their fast growth. At maturity, *Rorippa* and several other crucifers are much smaller than the other winter annual composites and would undoubtedly be competitively inferior to them had they been recruited at the same time. Thus, these winter annual crucifers successfully complete their life cycle by avoiding interference, instead of engaging in a losing contest with the bigger composites.

Single species dominance also occurs among the summer annuals. *Polygonum pensylvanicum* can form nearly monospecific patches in wet spots in fields, where it is competitively superior relative to other

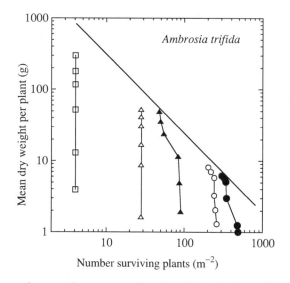

Fig. 7.2. Change in mean dry mass per plant in field populations of *Ambrosia trifida* of various densities during six harvests. Notice decline in numbers with time in high density populations.

early-successional plants (Pickett and Bazzaz 1978*b*). *Ambrosia trifida* can also be found in patches, especially on nutrient-rich soils. Its growth potential is remarkably high; as discussed in Chapter 6, it can overwhelm all other species present, forming essentially pure stands even when the initial seedling diversity is high. High-density populations show clear self-thinning during the growing season. But in low-density populations, all individuals survive (Fig. 7.2). Crowded individuals are usually single-stemmed, but individuals in low density branch profusely and attain great sizes. In these populations, the plants have high rates of leaf birth and death, a dense leaf canopy forms early, and the canopy is elevated with time by the birth of new leaves above the death of the old ones below (Fig. 7.3). This is assumed to be an economical investment of nitrogen as it is shunted from older leaves located in light limited environments deep in the canopy to newer, move active leaves in high light environment at the top of the canopy (see Bazzaz and Harper 1976, Field 1983). But under high density, there is both a death of whole plants and a decline in leaf birth and death rates. Despite differences in leaf life span among different cohorts (Fig. 7.4), leaves live longer in high density than in low density populations. Thus, population regulation occurs at both the whole individual level as well as at the leaf level. Competitive interactions among conspecifics can lead to other

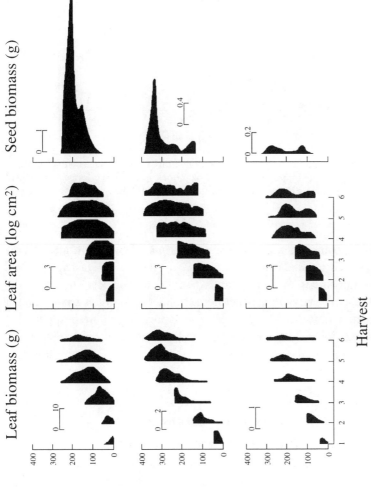

Fig. 7.3. Profiles of leaf biomass, total leaf area, and seed biomass in populations of the early-successional annual *Ambrosia trifida* growing in three densities in the field (modified from Abul-Fatih *et al.* 1979).

Plant–plant interaction

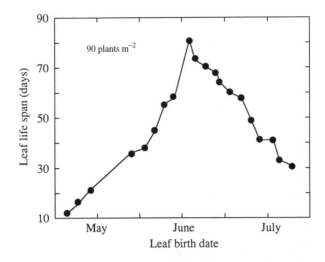

Fig. 7.4. Leaf life span in relation to time of leaf birth within the growing season of field populations of *Ambrosia trifida*. Leaves born in June live longest (modified from Abul Fatih and Bazzaz 1980).

changes in plant function. In *Ambrosia trifida* (Abul-Fatih and Bazzaz 1980) as well as *Ambrosia artemisiifolia* (Ackerly and Jasieński 1990), smaller individuals in dense stands tend to be females, whereas taller individuals are bisexual. Furthermore, smaller individuals have a higher reproductive allocation (RA) to seed than do taller individuals. Apparently, smaller individuals allocate their reproductive material only to seeds and take advantage of their taller neighbors for the pollen that rains down on the subordinate plants.

Interactions between species guilds
Winter annuals versus summer annuals
In many temperate successions, two groups of colonizing annual species are recognized: the winter and summer annuals (e.g. Oosting 1942, Bard 1952, Quarterman 1957, and Bazzaz 1968 in the USA; Numata and Yamai 1955; and Numata 1990 in Japan; Bornkamm, 1986 and Ellenberg 1986 in Central Europe). These two groups of annuals can replace each other, depending on the time of soil disturbance after field abandonment. When the soil is disturbed in the spring, the winter annual rosettes recruited in the fall are largely destroyed and the summer annuals predominate. Seedlings

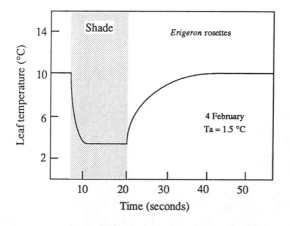

Fig. 7.5. Leaf temperature of *Erigeron canadensis* rosettes in winter can be significantly higher than air temperature, except when they are in the shade (modified from Regehr and Bazzaz 1976). Leaf temperature in the sun is near optimal for photosynthesis.

of the winter annuals may emerge in huge numbers under the canopy of the summer annuals. However, the majority of these seedlings die. Those that survive remain suppressed under the summer annuals, then grow in the following fall and winter, and dominate the fields during the second spring. Thus, the dominance and suppression of the winter and summer annuals does not result from differences in innate competitive abilities, but from the timing of disturbance in the fields, that changes competitive hierarchies. Either group of species could suppress the other, depending on disturbance time. The nature of interactions between the two major winter annuals (*Erigeron annuus* and *Erigeron canadensis*) and the archetypal, widespread summer annual (*Ambrosia artemisiifolia*) can be used as a model to illustrate the nature of interactions among winter and summer annuals.

Because they are located near the ground in the zone of low wind speed, rosettes of the winter annuals can have significantly higher leaf temperatures than that of the surrounding winter air on sunny days in winter (Fig. 7.5). Likewise, because they are able to seasonally adjust their photosynthetic responses to temperature, winter annuals can fix appreciable amounts of carbon during sunny, calm winter days (Fig. 7.6). Therefore, their rosettes can grow considerably before bolting in the spring. In this case, the winter annuals preempt space and associated resources, and suppress the newly emerging summer annuals in the spring. It is interesting to note, however, that the suppression of the summer annuals by the winter annuals results in

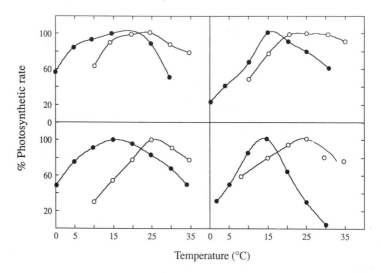

Fig. 7.6. Seasonal shifts in photosynthetic response to air temperature (acclimation) in the winter annuals *Rorippa sessiliflora* (top left), *Lactuca scariola* (top right), *Erigeron canadensis* (bottom left), and *Erigeron annuus* (bottom right); (●) photosynthesis in winter, (○) photosynthesis in summer (modified from Regehr and Bazzaz 1976).

much reduced growth but little mortality, especially in the most prominent of the summer annuals, *Ambrosia artemisiifolia*. In the field, removal experiments, modification of the physical environment, and augmentation of resources, especially nutrients and water, revealed much about the complex nature of interactions among these two groups of plants (Raynal and Bazzaz 1975b).

Augmentation of water and nutrient resources did not cause a shift in the relative competitive superiority of the winter annuals (Fig. 7.7); all plants grew larger but the relative growth of the species did not change. Progressively shading *Ambrosia* in field plots to simulate the influence of the winter annuals led to a reduction in its height growth. The use of filtered light to decrease red/far-red ratios greatly reduced the height growth of *Ambrosia* and caused an almost complete failure to produce seed. Thus, the suppression of the summer annuals by the winter annuals is largely caused by a reduced red/far-red ratio and decreasing light intensity (Fig. 7.8). The most effective treatment in releasing the growth of *Ambrosia* was plowing in the spring, which resulted in the near complete elimination of the winter annuals. Hand removal of the winter annuals also increased *Ambrosia*

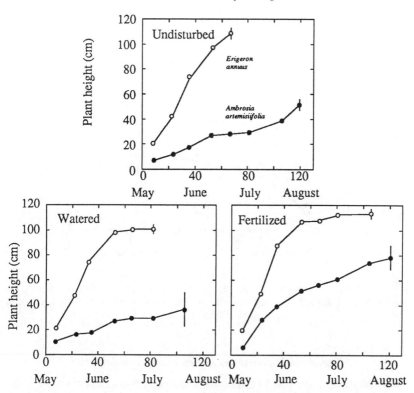

Fig. 7.7. The responses of the annuals to resource augmentation (water or nutrients). No change occurs in the relative position of the summer annual *Ambrosia artemisiifolia* in relation to the winter annual *Erigeron* (modified from Raynal and Bazzaz 1975*b*).

growth but not to the same extent as plowing (Fig. 7.9(*a*),(*b*)). While the density of *Ambrosia* in response to fertilizer treatment changed somewhat throughout the season, it declined drastically with the addition of water (Fig. 7.10). Seed production of *Ambrosia* was also increased by fertilizer and was completely inhibited by the addition of water. Augmentation of required resources caused different responses depending on the resource added. Fertilizer application resulted in higher growth and fecundity, and water addition resulted in death and the reduction of fecundity of the survivors. Furthermore, the addition of resources to alleviate competition caused two different outcomes: a positive response at the individual plant level and a negative response at the population level.

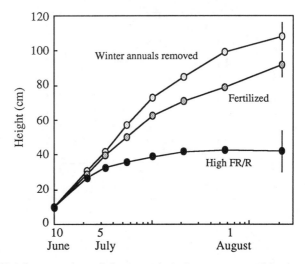

Fig. 7.8. Height extension of the annual *Ambrosia artemisiifolia* in relation to fertilizer addition, removal of the competing winter annual canopy above, or imposition of a high far-red/red ratio by the use of selective filters in the field (data from Raynal and Bazzaz 1975*b*).

The annuals versus the herbaceous perennial Aster

As we have seen, the guild of annuals is followed by perennial composites, especially *Aster* and *Solidago*. *Aster* usually attains prominence in successional fields in the second or third year. Recruitment of *Aster* occurs both in the first and second years after disturbance, but declines later in succession. Seeds may germinate in the fall or spring, and seedlings of this species grow slowly relative to the annuals. *Aster* commonly remains as small individuals under the canopy of large annuals in the first year; it then bolts and reproduces during the second year.

The fate of suppressed *Aster* individuals in the first year is determined by the identity of the annual competitors. For example, *Chenopodium album* and *Polygonum pensylvanicum* cause little *Aster* mortality, while equal densities of *Amaranthus retroflexus*, and especially *Ambrosia artemisiifolia*, cause much mortality. Also, growth response of *Aster* is dependent on the identity of its neighbors within a wide range of size, with *Ambrosia* as the most effective suppressor and *Amaranthus* as the least. Our work on this species has shown that: (1) *Aster* tolerates resource limitation imposed by the much larger annuals in the first year by remaining as unbolted rosettes in the understory; (2) rosettes can gain carbon and other resources even in this shaded situation by being able to photosynthesize at low light and by

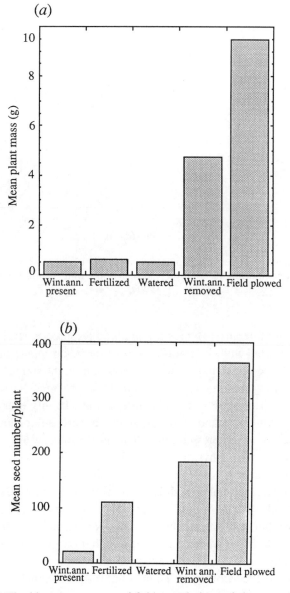

Fig. 7.9. (a) The biomass response of field populations of the annual *Ambrosia artemisiifolia* to treatments including hand removal of the winter annuals, the addition of fertilizer or water, and plowing that destroys the winter annual rosettes (modified from Bazzaz 1990a). (b) Seed production in *Ambrosia* in relation to removal of the competing winter annual or resource augmentation.

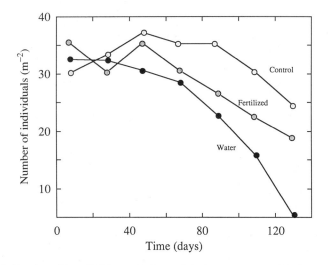

Fig. 7.10. Survivorship of field populations of *Ambrosia artemisiifolia* in response to resource augmentation (the addition of fertilizers and water) (modified from Bazzaz 1990*a*).

being drought tolerant; (3) *Aster* accumulates resources early in the spring and late in the fall while its annual competitors are dormant; and (4) *Aster* possesses the perennial habit which ensures that, at the start of the second or third growing season, older *Aster* individuals are much larger than the annuals, which must re-start as small seedlings. In this situation, *Aster* predominates in the second or third year and the early-successional annuals may be present as suppressed individuals in the understory.

Some of the same physiological attributes, however, may themselves preclude the continued success of *Aster* in the fields. The fact that *Aster* seeds germinate best under high levels of irradiance and moderately high nitrogen levels (Peterson and Bazzaz 1978), conditions associated with disturbance that removes vegetation, dictates that later *Aster* recruitment from seed under *Aster* stands is reduced and that the slow growth of young *Aster* seedlings results in their ultimate elimination from established *Aster* stands, despite the fact that the plants can produce as much as 0.5 million seeds per square meter.

Solidago *versus* Aster

Solidago altissima, *Solidago canadensis* and *Andropogon virginicus* usually replace *Aster* in the midwest, the east, and the southeastern part of the

North American deciduous forest respectively. Moreover, *S. canadensis* and *S. altissima* are now found in many oldfields in Europe and Japan and are becoming major plants in successional fields. The growth of *Solidago* in disturbed ground in Japan is particularly interesting, as individuals can attain much greater heights relative to plants in the eastern United States. *Solidago* and *Aster* differ in a number of physiological and morphological traits that influence their interactions (Schmid and Bazzaz 1994). *Aster* has many branches which bear many small leaves, while *Solidago* ramets are single-stemmed and carry larger leaves. *Aster* genets are not as compact as those of *Solidago* because they have longer rhizomes. Thus, in *Solidago*, sister ramets compete for light. In contrast, the distantly spaced ramets of *Aster* may be penetrated by ramets of other individuals or other species. Light penetration is higher in *Aster* canopies than in *Solidago* canopies. The two species do not differ greatly in their photosynthetic rates. It is likely that because of the differences in light penetration, seedlings of *Solidago* can emerge under *Aster* stands, but the reverse is less common. Furthermore, clonal integration, which is stronger in *Solidago* than in *Aster*, may confer some superiority to the former relative to the latter. This high integration in *Solidago* may lead to the success of genets, even when parts of them are located in less favorable habitats (Hartnett and Bazzaz 1985*a*). *Solidago* has lower leaf turnover and longer-lasting leaves with a higher nitrogen content than *Aster*. *Solidago* also rapidly shifts carbon allocation as light conditions change. Carbon-14 labeling experiments show that *Aster* shunts ^{14}C toward upper leaves and nodes, whereas *Solidago* shunts ^{14}C more equally to both upper and lower nodes. Because relatively more photosynthate can be shunted to underground parts of *Solidago*, these resources help maintain *Solidago* in place longer than *Aster* (Fig. 7.11). After only a short time, the *Aster* genets usually fragment, while the penetrating *Solidago* individuals grow by forming many connected, long lasting, physiologically integrated ramets in compact, difficult to penetrate genets (Schmid and Bazzaz 1994).

It is puzzling that *Solidago* rosettes appear in oldfields two to three years later than those of *Aster*. There is no reason to suspect that dispersal could account for this. Seeds of both species are well suited for wind dispersal and do not differ greatly in size. Furthermore, under controlled conditions with ample nutrients, light, and moisture, seeds of both *Solidago* and *Aster* germinate readily and produce vigorous rosettes which bolt and flower in one growing season.

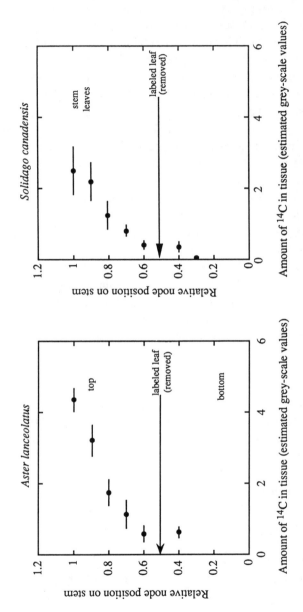

Fig. 7.11. ^{14}C distribution in leaves of different positions in the canopy of *Aster* and *Solidago* stands (modified from Schmid and Bazzaz 1994).

Solidago *versus trees*

Seed dispersal, efficiency of resource capture and use, and tolerance of (in some cases, preference for) low resource levels especially light levels, may interact to produce the observed changes in community structure from early to mid- to late succession, where, in many geographic locations, the herbaceous perennial community is replaced by shrubs and trees. Differences in biomass allocation patterns play a major role in the replacement of *Solidago* by the invading trees. Complete loss of aboveground parts during the dormant season in *Solidago*, and the rebuilding of that biomass from underground rhizomes during the following growing season, is a disadvantage when *Solidago* confronts trees. Tree seedlings and saplings drop only their leaves, retaining their stems and dormant buds. As a result, *Solidago* starts its growing seasons at the ground level, while in the absence of herbivory and severe dieback, the invading tree seedlings and saplings start successive years' growth from the position of their terminal buds in the previous year. With a more or less fixed canopy height, *Solidago* grows largely in two dimensions by clonal spread, while the trees grow in three dimensions and overtop *Solidago* after a few seasons.

Solidago invests much in clonal expansion; new ramets are very dependent upon underground reserves and their older sister ramets for their early support (Hartnett and Bazzaz 1985*b*). Early in life, tree seedlings usually allocate a substantial portion of their accrued biomass to belowground growth, developing more extensive and more deeply penetrating root systems than *Solidago*. *Solidago* may even afford invading tree seedlings shade conducive to early establishment and protection from high wind, low humidity, and large mammalian herbivores such as deer (Burns and Honkala 1990).

Although there are clear differences among tree species in shade tolerance and their associated morphological, physiological, and architectural traits, which would suggest the relative importance of the competitive interactions discussed above in determining successional trajectories, entry into oldfields of particular tree species is strongly influenced by dispersal patterns and chance. Horn (1976) suggests that, in some successions, if propagules are available, patchy disturbance can create circumstances where any tree species can replace any other species, irrespective of their level of shade tolerance. Thus, competition with *Solidago*, though significant once species are established, is not determinant of succession patterns.

Is competitive competence a fixed character in early-successional plants?
There is strong evidence that the intensity and outcome of plant–plant interactions among early-successional species depend on the level of environmental controllers (such as temperature) and resources (such as light, nutrients, water, and CO_2) experienced in the field. Only in limited cases can one species be said to be competitively superior to another in all conditions occurring in the field. Especially among annuals, which must be recruited every year, conditions during the time of germination and emergence can play a major role in defining competitive hierarchies. Competitive shifts commonly occur among the annuals along gradients of soil moisture (Pickett and Bazzaz 1978*b*), nutrient levels (Parrish and Bazzaz 1982*c*), and atmospheric CO_2 concentrations (Bazzaz and Garbutt 1988). Different resource levels in various patches in the field and year-to-year variation in the weather can generate different competitive hierarchies. Thus, in many situations the search for consistent competitive hierarchies will continue to be elusive. Therefore, inconsistencies among findings and interpretations by various investigators will, unfortunately, continue to be encountered unless we define the conditions under which the competitive competence of a species (or genotype) is evaluated.

Although the role of competition in successional change was recognized very early in the development of successional theory, much of the work on the mechanisms of plant–plant interactions has been concentrated on herbaceous plants and little is known about tree species. Furthermore, our knowledge is very limited for plant–plant interactions in habitats acutely or chronically dominated by shortages, excesses, or great imbalances in resource availability. Rapid anthropogenic change in biosphere–atmosphere interactions, such as the rising atmosphere CO_2 concentrations, may also strongly influence plant–plant interactions, including those that shape successional patterns (Bazzaz and McConnaughay 1992). Inputs may interact with each other and with the soil environment to produce enormous changes in the outcome of plant–plant interactions in successional environments.

8

Interaction and the evolution of response breadths and niches

The concepts of coexistence of species and of species diversity have been central to the understanding of community organization (May and MacArthur 1972, Whittaker and Levin 1976, and see Diamond and Case 1986). Theoretical and experimental work suggests that species must have different resource requirements in order for them to coexist in a community (May 1973, Bergh and Braakhekke 1978). Therefore, coexisting species are assumed to occupy different 'niches' and are allowed only a minimum amount of niche overlap. Despite their presumed importance, however, several decades of research about niches have not produced a widely accepted unified concept, especially for plants. Much confusion has resulted from a lack of clear, generally accepted definitions. In fact, some authors (e.g. Connell 1978, Huston 1979, Silvertown and Law 1987, Harper 1988) have questioned the utility of the niche concept for plants. These authors argue that because all plants require the same resources from their environments – i.e. light, nutrients, water, and CO_2 – they cannot occupy distinctly different niches. In contrast, Tilman (1988) argues that this similarity in resource requirement does not necessarily limit the diversity of species in a community. He proposes that the differentiation among species in the ratios of nutrients required for growth can account for their coexistence. Tilman (1986) considers that soil nutrients and light form a natural complex gradient. Each plant has a certain region along this gradient in which it is a superior competitor. He feels that such gradients may have been major axes of differentiation in the evolution of early plants and may help explain the life history patterns of current plants.

The 'niche' has been defined as:

1. The habitat in which a species makes its living (Grinnell 1928).
2. The role of the species in the biological environment: what it does and how it lives (Elton 1927).

147

3. The hyperspace–hypervolume, defined by several environmental axes, in which the species can exist indefinitely: 'the n-dimensional niche' (Hutchinson 1957).
4. The way a species is specialized within a community, its position in space and time, and its functional relationship to other species in the community (Whittaker 1972).

Whittaker's conception seems to combine elements of the previous definitions and may be the most operationally suitable for plants. It includes notions of the niche as a 'function' and as a 'habitat.' Whittaker (1975) proposed that plants can differ in their use of resources, time of activity, vertical and horizontal location of shoots and roots, etc., and that these differences themselves are axes of the plant niche. Cody (1986) emphasized life form differences and strategies of light interception and nutrient uptake as promoters of coexistence in plants. Grubb (1977) recognized four components of the plant niche: the habitat niche, the life-form niche, the phenological niche, and the regeneration niche. He suggested that plant coexistence could be promoted by differences among species in any or all of the four, but emphasized the role of the regeneration niche.

How large must the niche difference be to permit species coexistence? Theoretical analysis (e.g. May 1974, Newman 1982) produce a range of minimum values of overlap required for coexistence. However, in plants severe limitation can be imposed on the evolution of niche differentiation. As mentioned earlier, unlike food resources for animals, critical plant resources such as light, water, and nutrients are usually presented to the plants in continuous rather than discrete packages. Plants also compete strongly for space itself (Connell 1980, Yodzis 1989) above and below the ground, irrespective of the amount of resources in it. Furthermore, the intriguing finding that members of a plant community can be physically connected to each other by mycorrhizal fungi and therefore actively share some resources (e.g. Chiariello *et al.* 1982) would reduce the potential for niche divergence on soil resource axes. In contrast, biological interactions such as predation, herbivory, and pollination are strong in most plant communities (Connell 1971, Janzen 1981), and plants may diverge more on these biological axes than on physical axes. Thus, the similarity in resource requirements and competition for space can put severe limitations on evolutionary niche differentiation in plants.

A plant niche is not just its 'function' or its 'habitat.' In fact, if coexistence among species is a critical requirement for the evolution of their niches, then location of a species on differing parts of habitat gradients (or having their activities occur at different times, for example differences in flowering phenology) may have limited meaning for 'coexistence' and therefore for niche. 'Niche' *per se* cannot be inferred from mere habitat separation, unless it is proven to be the result of past reciprocal selection. A plant niche is best defined in terms of the interaction of the plant with its environment. How broad is the response of a species to various critical resource axes, and what is the volume of this response along these axes? Thus, the concept of 'niche' resembles that of the n-dimensional niche of Hutchinson and is allied to the classical notions of 'ecological limits' or 'tolerance limits.'

In defining the plant niche, we must clearly distinguish between the terms *coexistence* and *co-occurrence*, which are sometimes used to mean the same thing. Despite their common use, these concepts are still operationally vague; they are not rigorously defined, especially with regard to time scale. It is not clear, for example, how long two populations must remain together in a given location before one replaces the other or are both replaced by another population in the process of normal succession. And do they have to repeatedly reproduce in the same location in order for them to coexist? 'Coexistence' suggests long-term co-presence, which may involve reciprocal selection, while 'co-occurrence' suggests short-term co-presence, which is disrupted by normal processes of community change. In this case, co-occurrence may involve populations that can be very different in their life histories, or very similar, since there is no requirement for niche differentiation.

The limited experimental work and theoretical considerations that have focused on precisely defining the plant niche show that: (1) there are many physical and biological factors to which the plants respond; (2) a plant's response is rarely, if ever, equitable across the environmental gradients that the species is likely to encounter in its environment; (3) it is unlikely that many of these resource axes are independent of or orthogonal to each other; and (4) the response of an individual on any axis may be greatly influenced by the identity of its neighbors and the levels of other critical resources. Thus, in order for the concept of the niche to be useful in understanding plant responses and their evolution, a new definition must be sought that would take into consideration the particular attributes of plants that differentiate them from animals (since the original concept of niche was formulated for animals).

Using the phytocentric view of plant niches can give us a way to express the nature of response of individual populations and species to their major factors in their environment. It allows an accurate quantification of, and comparison between, interacting species and the role of these interactions in community organization and succession. While these analyses can inform issues of community structure, they may not be of great help in elucidating coexistence in communities, because the minimum allowable overlap of niches between species that is sufficient for coexistence remains unknown and may differ from one community to another. Of course, the concept of coexistence in plant communities, especially in successional ones, remains rather fuzzy and without clear definition.

How to determine a plant's niche

If the critical factors to which a plant responds are identified, and its response on a range of these factors is experimentally determined, or if the patch types on which the species is found are known, identified, and ordered in a gradient, either graphically or mathematically using multivariate techniques, then a space (or volume) of response would delineate the niche of the species (Fig. 8.1). Without competition, this would be its tolerance limits or Hutchinson's 'fundamental niche,' and with competition, it would be its ecological distribution or Hutchinson's 'realized niche.' The fundamental niche of a species is obtained when resources are uncontested, while the realized niche depends on the identity (and the biomass) of the neighbors and how they modify the resources and conditions in the habitat. In this view, then, the niche is delineated by the plant by the way in which it responds to its environment. It assumes nothing about coexistence or about the length of 'co-presence' of species in a community. Species, populations, or individuals can have niches. That is, a plant niche may be defined as the pattern of response of an individual, a population, or a species to the critical physical and biological gradients of its environment (Bazzaz 1987). Therefore, for plants 'niche volume' is more appropriate than 'niche breadth' or 'niche width.' Response breadth is related to tolerance limits on a single factor, while niche volume is related to ecological amplitude, which encompasses several critical factors of the environment.

The terms *niche separation* and *niche differentiation* have sometimes been used interchangeably. *Niche differentiation* is best used to refer to differential resource use or response that results from long-term, consistent competitive interactions between or among species in a community where the species populations act as selective agents on each other. It results from coevolutionary resource use displacement. *Niche separation* refers to the differences in resource use patterns among species that may or may not involve niche differentiation. These differences may have evolved indepen-

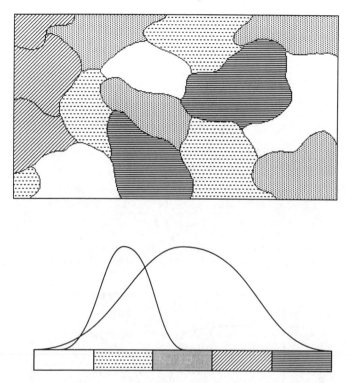

Fig. 8.1. Hypothetical range of environmental patches in a field, the arrangement of these patches in ordered resource gradients, and the distribution of 'specialist' and 'generalist' plants. The first is found in three patch types and the second in all patch types.

dently, as a result of competitive interactions with other species in another community. The former is extremely difficult to prove and the latter is easier to detect.

Many plant species of early-successional communities are introduced from other locations. Therefore, it is unlikely that they have experienced consistent and strong interactions resulting in reciprocal selection and niche differentiation. Instead they can coexist with other species in the community because of niche separation. For example, the composites *Erigeron annuus* and *Lactuca scariola* flower during June and July in one-year successional fields. They attract the same insect visitors (mostly syrphid flies and halictid bees). *Lactuca* flowers open early in the morning, whereas *Erigeron* flowers open in midmorning after most *Lactuca* flowers have closed (Fig. 8.2). Direct observations in the field of change in pollen loads of the pollinators indicate that these pollinators switch to *Erigeron*

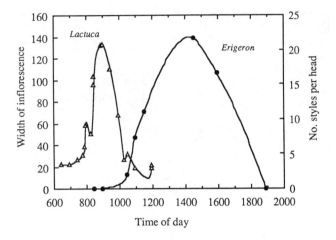

Fig. 8.2. Daily time of flower opening in the composites *Lactuca scariola* and *Erigeron annuus* in early-successional fields (modified from Parrish and Bazzaz 1978).

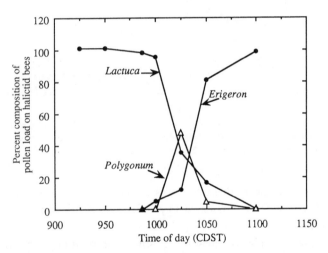

Fig. 8.3. Percentage of pollen load of *Lactuca*, *Polygonum*, and *Erigeron* on insect pollinators in the annual community (modified from Parrish and Bazzaz 1978).

from *Lactuca* after the flowers of *Lactuca* close. Some pollinators, for example *Halictus ligatus*, opportunistically foraged on *Polygonum pensyl-vanicum* pollen during the transition between peak flowering of the two composites (Fig. 8.3). *Lactuca* is an introduced species from Europe. In its native habitat it co-occurs with *Senecio* species, which resemble *Erigeron* in their flowering behavior (Parrish and Bazzaz 1978), and may have evolved its daily flowering strategies in response to *Senecio*. Therefore *Lactuca* was preadapted to enter the early-successional community in the midwestern United States.

Niches in successional ecosystems

Odum (1969) was the first to make explicit predictions about the differences in response (niche) breadth of early- and late-successional plants, primarily based on the r-K continuum of strategies of MacArthur and Wilson (1967). He proposed that early-successional plants have broader and more overlapping niches than do late-successional plants. These predictions are intimately tied to assumptions about growth rates of populations and strength of competitive interactions among species in early- and late-successional communities and their evolutionary outcome (Chapter 6). We investigated the patterns of plant niches in successional communities by developing and experimentally testing a series of hierarchical predictions for plants in these communities. Our hypotheses are based largely on differences in disturbance regimes and the nature of changes in the environment commonly experienced by plants occupying different successional positions during ecosystem recovery. We made the following predictions:

1. Early-successional plants, which occupy repeatedly disturbed habitats with variable and changing environments, should have broad responses on major resource gradients and, therefore, should have large niches.
2. Because many early-successional habitats change greatly within a season, from being open and exposed early during recruitment to being closed later, various life stages of their component plants experience different environments and may exhibit varied adaptations to these environments. Their niches will vary ontogenetically.
3. Because early-successional plants have large niches, they overlap greatly on resource gradients.
4. Communities made up of plants with large niches are expected to have low species diversity, as each species occupies a large portion of the available niche space, thus allowing for limited species packing.
5. Due to a high degree of niche overlap among early-successional plants, these species are expected to experience a high degree of competition. Some of them therefore undergo much biomass reduction in situations of severe crowding.
6. Because species of different phylogenetic affinities co-occur in communities, the response of early-successional plants, while generally broad, should

Table 8.1. *Response breadths and proportional similarities of plants from early and late succession on several individual niche axes*

	Early succession	Late succession
(a) Mean response breadth (calculated as Levins' *B*) of all species in experimental assemblage		
Herbaceous		
Underground space	0.71	0.29
Pollinators	0.20	0.16
Nutrients	0.77	0.70
Trees		
Nutrients	0.91	0.82
Moisture	0.89	0.65
(b) Mean proportional similarity (an estimate of overlap in response) of all species in experimental assemblage		
Herbaceous		
Underground space	0.68	0.43
Pollinators	0.31	0.19
Nutrients	0.85	0.83
Trees		
Nutrients	0.94	0.83
Moisture	0.82	0.70

Source: Bazzaz 1987.

not be the same for all species on the same gradient nor for any species on all gradients. That is, there cannot be complete ecological convergence among the major species.

7. Because early-successional plants have broad responses to resource gradients (relative to late-successional plants), they will be similar to each other in their interactions and therefore show a high degree of equivalency as neighbors.

To test these predictions, we studied the response breadth of species from early, mid-, and late succession, including annual and perennial herbs and trees. We used several gradients in these studies, including nutrients, moisture, light, temperature, CO_2, diversity of competitive neighborhoods, and pollinators. We assessed response in most cases on one gradient at a time, and in some instances, we calculated response on two crossed dimensions (Table 8.1). We also considered the separation of response along a time axis and compared the response of species at different times during their ontogeny. The results were examined at within-community and between-community levels.

Table 8.2. *Response breadth (Levins' B) of the five dominant early-successional annuals on four environmental gradients*

Species	Nutrients	Light	Moisture	Temp.
Abutilon theophrasti	0.60	0.71	0.88	0.62
Amaranthus retroflexus	0.69	0.83	0.49	0.78
Ambrosia artemisiifolia	0.72	0.91	0.76	0.95
Chenopodium album	0.81	0.88	0.87	0.54
Polygonum pensylvanicum	0.82	0.90	0.99	0.91

On a within-community level, the species differed among each other on the same gradient and also differed on different gradients (Table 8.2). On the nutrient gradient, the species differed slightly in their germination breadth but significantly in their survivorship, with the early-successional species surviving on a wider range of nutrients than did the late-successional species (Fig. 8.4). Early-successional annuals also differed among themselves in the timing and placement or deployment of root growth in the soil. For example, *Ambrosia* tended to grow most roots early in the season, while *Setaria* grew them late in the season (Fig. 8.5) along all gradients. The early-successional communities consistently had more overlapping or broader responses than did late-successional communities (Table 8.1).

Differences among early-successional plants in the placement of roots were observed in the field as well. *Setaria faberi* has a fibrous root system that tends to be concentrated at the upper parts of soil profile. *Abutilon theophrasti* locates many of its roots in the middle region, while the roots of *Polygonum pensylvanicum* extend down to the water table, deep in the soil profile. The location of the root system may allow access to different soil and water resources and may influence the pattern of the plants' daily water potential and the way they respond during periods of drought and after heavy rains. Photosynthetic response to leaf water potential differed among the three species (Fig. 8.6) such that all neighboring individuals, irrespective of their identity, could operate near maximum, despite the great differences among them in leaf water potential (Wieland and Bazzaz 1975).

Of particular interest were the studies on competition for pollinators, since biological resources are perhaps very important for niche differentiation, yet are less commonly considered in studies of plant competition. Studies of competition for pollinators among species in both early- and late-successional herbaceous plants revealed differences among them. In both communities, flowering times are somewhat clumped in three flowering assemblages: spring/early summer, midsummer, and late summer/early fall. There were

156 *Interaction and evolution*

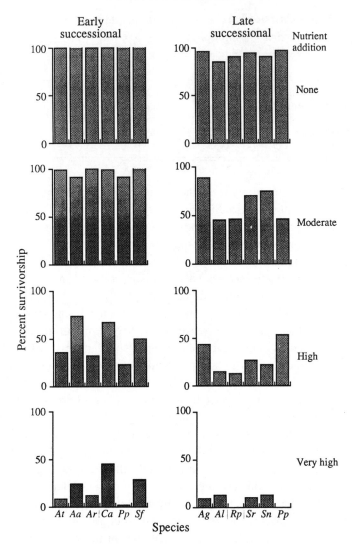

Fig. 8.4. Survivorship on a nutrient gradient of species from an annual early-successional community (left) and a late-successional grassland community (data from Parrish and Bazzaz 1982c). *At, Abutilon theophrasti; Aa, Ambrosia artemisiifolia; Ar, Amaranthus retroflexus; Ca, Chenopodium album; Pp, Polygonum pensylvanicum; Sf, Setaria faberi; Ag, Andropogon gerardii; Al, Aster laevis; Rp, Ratibida pinnata; Sr, Solidago rigida; Sn, Sorghastrum nutans; Pp, Petalostemum purpureum.*

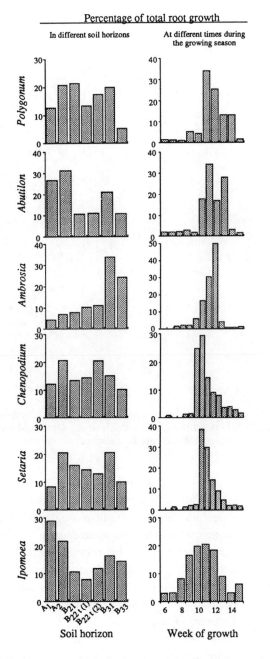

Fig. 8.5. Difference among co-occurring annuals from an early-successional community in the location in soil horizons and timing of root growth within a single growing season (modified from Parrish and Bazzaz 1976).

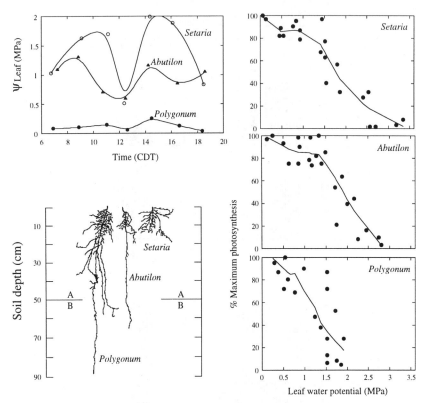

Fig. 8.6. Location of roots in a deep soil of the co-occurring *Setaria, Abutilon*, and *Polygonum*; leaf water potential of individuals in the same field location; and the response of photosynthetic rate to a decline of leaf water potential.

differences among species in the two communities, both in the identity of pollinators and in their daily and seasonal times of visits. In general, there was broader response and more overlap among species in the annual community than among species in the late-successional grassland community. The species from the late-successional community also showed a higher degree of specialization for pollinators than did the annuals. In both communities, however, the overlap in response was lower than what it would be for a random community, suggesting some degree of niche separation among species (Table 8.1).

The results of all these studies on niche relations in early- and late-successional communities have been remarkably consistent with our predictions, despite differences in response among species within guilds and across gradients. In every case, early-successional species had broader and

Single species   Community

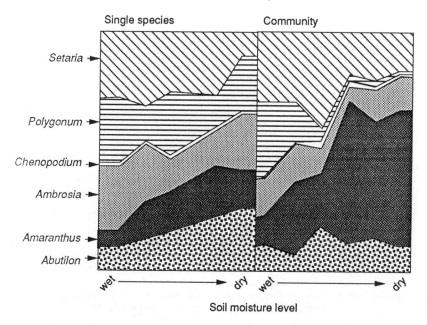

Soil moisture level

Fig. 8.7. Differential response of the summer annuals on a moisture gradient grown as individuals and in a community (data from Pickett and Bazzaz 1978*b*).

more overlapping responses than did late-successional species, whether the comparisons were made among herbs or among trees. Although early-successional plants generally have broad responses, the differences among them can become accentuated when the species are grown in competition with other members of the same community. There can be much reduction in their response surfaces. For example, in competition on a moisture gradient, *Polygonum pensylvanicum* shifts strongly toward the wet end, while *Abutilon theophrasti* shifts slightly toward the dry end (Fig. 8.7). These shifts may contribute to co-presence in this community, as different species perform better on different parts of the gradients and occupy different patches in the field. In the early-successional annuals, there are many instances where the response of suppressed individuals on a gradient showed a decoupling of growth from reproduction (Fig. 8.8). The extent of this decoupling is not known for plants in general, despite the fact that it may be critical to the persistence of populations and to the evolution of response breadth and niches in plants. In evolutionary terms, the shape of the 'fecundity surface' may be more critical than that of total biomass. It is also more critical to community dynamics.

What do these patterns of response mean in terms of plant–plant

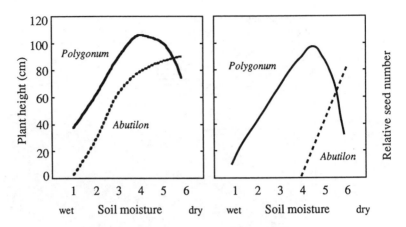

Fig. 8.8. Differential response of *Polygonum* and *Abutilon* in competition on a moisture gradient. *Polygonum* is taller than *Abutilon* on the wet end and shorter on the dry end of the gradient. There is also a clear decoupling of reproduction from vegetative growth in *Abutilon*, but little in *Polygonum*. *Abutilon* produces vegetative biomass but no seed on the wet and moist parts of the moisture gradient in competition (data from Pickett and Bazzaz 1978*b*).

interactions in changing environments? Broad response along environmental gradients and a higher degree of overlap in response can result in stronger interactions in areas of severe overlap. Thus, when early-successional species grow together in dense stands, competitive interactions among them ought to be very strong. In contrast, because late-successional plants have relatively narrow responses, and their overlaps are potentially small, competition among them should be less intense. Theoretically, intensely competing species with much overlap in resource use may diverge in their response over evolutionary time, resulting in more resource sharing, reduced competition, more niche differentiation, greater species packing, and higher species diversity. Evidence for the existence of such a response is only circumstantial, and the effect of competition on niche separation along several resource axes has only been inferred. Ecologists have long argued about the importance of competition in the shaping of response breadth and in community organization (see Connell 1980, 1983). While the differences in response among species may contribute to their coexistence, there is little experimental evidence that competition actually leads to niche differentiation in plants.

We have experimentally compared the strength of competition among plants in early- and late-successional communities by testing the ability of species in each group to obtain resources and to use them for growth and

Table 8.3. *Mean biomass per individual competing in conspecific and in heterospecific pairs in early-successional annual and in late-successional prairie communities*

	Biomass (g)	
	Individual in conspecific pair	Individual in heterospecific pair
Annuals	1.461	1.092
Prairie species	1.386	1.616
	ns	$p < 0.001$

reproduction (Parrish and Bazzaz, 1982*a*). We grew plants from these communities singularly, in con- and heterospecific pairs, and in within-community pure and mixed stands. The biomass of individuals in heterospecific pairs relative to the biomass in conspecific pairs was significantly lower for early-successional than for late-successional plants (Table 8.3). Most of the pairwise combinations of the early-successional species had relative biomass ratios of less than one; i.e. both members of a pair accumulated less biomass in heterospecific than in conspecific competition. In contrast, late-successional species had relative biomass ratios above or below one in almost equal numbers of combinations. Apparently, competitive interactions were more prevalent in the early-successional community, resulting in a decreased ability of both species to obtain resources and to grow. The biomass of mixed species pairs, divided by the biomass of individual species grown singularly, showed that the early-successional species experienced significantly more reduction in total biomass than did the late-successional species. Furthermore, the species of the late-successional community had a higher relative yield when grown in mixed stands than when grown in pure stands. Total yield in mixed stands was only 7% higher than the yield in all pure stands of the early-successional community, but was 34% higher in the late-successional community. It appears that selection to reduce competition could have been more important in the evolution of late-successional species than for the early-successional ones. Apparently, despite strong competitive interactions among species of the early-successional community, there has been little niche differentiation, because competitive interactions are inconsistent and do not occur often enough. It may be more selectively advantageous to be a generalist and a poor competitor (the tragedy of the jack of all trades) in early colonization and succession. In this and many other annual communities of early succession, the presence of fugitive/ruderal introduced species may confound competitive interactions. Thus, there may not be enough time for the reciprocal selection to occur. Also, long-term persistence of seed banks of early-successional plants may contribute to the inconsistency of interactions.

It is expected that over some time selection will generally reduce the response breadth and niche volume of species in the community. However, this is not the case in the early-successional community. In fact, the overlap observed on a moisture gradient may approach the mean maximum similarity possible ($x = 0.66$). How, then, is the process of differentiation opposed in early-successional plants such that their responses on gradients remain broad and their niches large? We suggest that the unpredictability of habitat conditions and the continued change in the identity of neighbors at a site are strong enough to oppose niche divergence. Furthermore, the external determinants of neighborhood structures are more unpredictable in early-successional habitats. The timing, location, and magnitude of disturbance may be especially important. To test the degree of similarity in resource use in the early-successional community and to see if there is any divergence in resource use we performed a simulation (Pickett and Bazzaz 1978*b*). For a community of six species growing on a broad environmental gradient, the inputs of random values in a raw data matrix generated a mean overlap after many iterations of 0.68, which is not significantly different from the mean similarity calculated from species performance in an experimental mixed stand on a moisture gradient. This strongly suggests that there has not been any differentiation on this gradient.

Response shifts during plant ontogeny

Because plants are sessile, modular, and grow by the addition of more modules, they experience a changing environment during their lives by expanding into new locations in three-dimensional space. In exposed, early-successional habitats, most seeds germinate near the soil surface. As a result, the young seedlings are found in the portion of the environment that is likely to be the most variable. Temperature fluctuations are greatest at the soil surface and adjacent few centimeters. Soil moisture is also most variable near the surface, as it becomes wet during rainfall and dries out soon after. Carbon dioxide concentrations are usually highest near the soil surface (Chapter 4). It is expected, therefore, that relative to older, larger individuals, seedlings experience (and tolerate) a broader range of environmental variables. For example, seedlings and adults of *Ambrosia*

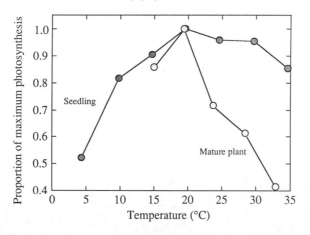

Fig. 8.9. Unacclimated relative response of seedlings and mature plant photosynthesis to air temperature in the pioneer species *Ambrosia artemisiifolia* (modified from Bazzaz 1974).

artemisiifolia grown in controlled environments under similar conditions differ in their photosynthesis response to temperature, with the seedlings having a much broader response than the adults (Fig. 8.9). In order to further investigate niche shifts with plant ontogeny, we compared seed germination, seedling biomass, and reproductive biomass of the early-successional annuals on gradients of moisture, nutrients, and temperature (Parrish and Bazzaz 1985a). The species, as well as the stages in the life cycle of individual species, differed on the various experimental gradients (Fig. 8.10). Not only did the plants responses change with age, but for moisture and nutrients, plant responses of the same stage were more similar to each other for different species than for different stages of the same species (Table 8.4). Apparently, selection on these gradients was so strong in early-successional habitats that it resulted in the convergence of responses of species of different taxonomic affinities that occupy similar habitats. Again, patchiness in distribution of these species in the field and seed longevity in the seed bank contribute to this convergence by reducing divergence due to consistent competition.

Structure of the population niche

The preceding discussion compared and contrasted response breadth and niche volume and their contribution to species coexistence in plant communities. Families (progeny of an individual) and individuals within

164					Interaction and evolution

Table 8.4. *Number of pairwise comparisons in which species are most similar to (a) other species of the same age, or (b) other ages of the same species*

Gradient	(a)	(b)	p
Nutrient	27	9	<0.002
Moisture	33	3	<0.001
Temperature	21	15	>0.20
Light	4	6	>0.50

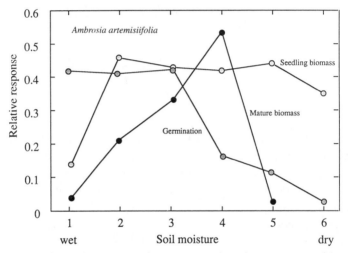

Fig. 8.10. Shifts with ontogeny in response of *Ambrosia artemisiifolia* to soil moisture (modified from Bazzaz 1987).

populations may also show differences in response to the environment, particularly in terms of fecundity. These differences could be critical to the genetic structure of populations in the field. In order to assess the potential for evolution, fecundity response breadths must be precisely quantified. It is critical to understand how these individuals respond to the range of environments they encounter; the range of their response breadths must be known. It has long been recognized that a population can consist of a set of individuals nearly equally broad in their responses to the environment; i.e. each can cover a range on the gradient close to that of the entire population. On the other extreme, each individual in the population can be highly specialized on a narrow range of the environmental gradient, showing only little overlap in response. In this case, the response of the population is much broader than the response of any individual. The first situation will

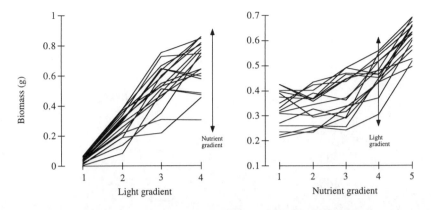

Fig. 8.11. The biomass of *Abutilon* families from a single field population on a light and a nutrient gradient (modified from Garbutt and Bazzaz 1987).

lead to a small between-genotype and a large within-genotype component of response, whereas the second will lead to the reverse: a high between-genotype and a low within-genotype response (Lewontin 1957, Roughgarden 1979, Pianka 1994).

It is expected that early-successional plants should show broad responses to their physical environment as individuals and as populations. However, analysis of *Abutilon theophrasti* families obtained from several individuals from one population in an early-successional field shows that the structure of the response was different along nutrient and light gradients (Garbutt and Bazzaz 1987). Furthermore, each population produced a combination of families with narrow and broad responses, particularly on the light gradient (Fig. 8.11). However, all families had a very broad range on the nutrient gradient. Despite the overall breadth of response performances, a few families differed greatly from the others. For example, two families (A and B) had very similar total seed production across the entire gradient, yet the distribution of seed weight among various states of the gradient was very different (Fig. 8.12). Family A produced seeds in all states except the first (low nutrients), and family B produced virtually all its seeds in state 5 (high nutrients). Thus, relative fecundity of these families differed greatly across various ranges of the gradient. Because of seed longevity in the soil, which greatly slows the response of selection (see Chapter 5), and since patchiness changes over time one genotype can not overwhelm another on different portions of the gradient in an early-successional stand.

Partitioning variation among individuals in a population can be useful in understanding the evolutionary history of the population. In order to

Interaction and evolution

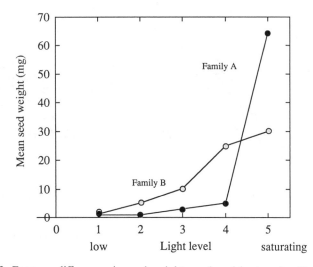

Fig. 8.12. Extreme differences in seed weight produced by two families from the same field population grown on a light gradient (modified from Garbutt and Bazzaz 1987). Both families produce the same amount of seeds along the entire gradient.

distinguish whether such variation represents multitude of specialized genetic variants or phenotypic plasticity of a few generalist genotypes, one needs to study growth variation of clonal copies of genotypes in common-garden, controlled environmental conditions. Seeds of *Abutilon theophrasti* and *Polygonum pensylvanicum* were randomly sampled in an early-successional field. Plants were propagated using an axillary bud enhancement technique. This technique begins with a single seed per genotype; the individual grown from this seed is then used to produce, through tissue culture, a clonal line of genetically identical daughter plants. We grew 27 genotypes of *Abutilon* and 25 genotypes of *Polygonum* under a variety of ecological conditions, encompassing environmental variation encountered by those species in their original location. Responses of the genotypes to each of the experimentally-imposed gradients can be expressed as norms of reaction.

Variation among genotypic reaction norms was pronounced only within certain ranges of the environmental gradients. For example, maximum divergence among *Abutilon* genotypes occurred at highest nutrient levels (Fig. 8.13), with genotypes responding similarly to less fertile soil conditions. Genotypes grown on the temperature gradient, on the other hand, maintained a relatively broad band of variation throughout the range of experimental conditions, with intermediate temperature levels promoting better growth.

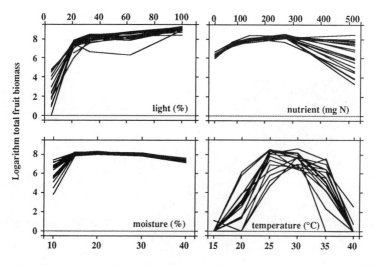

Fig. 8.13. The response of *Abutilon* genotypes from the same population to light, nutrient, soil moisture, and temperature gradients.

Type of response may also be not only genotype-specific, but also trait-specific. For example, individual *Polygonum* genotypes grown on the nutrient gradient differed in average biomass of fruits produced, but the trait itself was generally unresponsive to increased nutrient fertilization. Fruit number, however, exhibited a more complex response: some genotypes were not capable of benefiting from higher nutrient levels and their reaction norms remained flat. Other genotypes not only produced significantly more fruits at higher fertilization regimes, but also managed to overtake genotypes with higher fruit production at lower nutrient levels. This scenario represents an interesting evolutionary case of crossing norms of reaction, indicating that selective ranking of genotypes may change according to, in this case, edaphic conditions. To summarize, in both species all genotypes exhibited a generalist pattern of response, rather than specialize within a narrow segment of the experimental gradient.

It is critical also to emphasize the interactive nature of resource influences on the performance of genotypes. The growth responses on any one gradient may be significantly modified by the levels of other resources. Fully-crossed factorial experiments are needed to address this interaction. Therefore, we grew 25 genotypes of *Polygonum pensylvanicum* in the experimental garden under 20 combinations of five levels of light intensity and four levels of nutrient fertilization. One genotype (Fig. 8.14) had the highest fruit production at the combination of highest nutrient addition

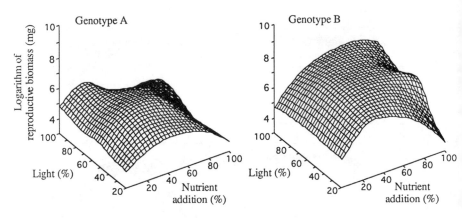

Fig. 8.14. Differential response of two genotypes of *Polygonum* on interactive
nutrients and light gradients.

and highest light intensity. Intermediate nutrient levels had a compensatory
effect, by improving fruit biomass production in both genotypes when they
were grown at lowest light levels.

Apomictic reproduction is not a common syndrome in successional
plants. Early-successional plants tend to be self-compatible, and late-
successional plants are generally outcrossing. However, the apomictic
species *Antennaria parlinii* can be found in large populations in mid-
successional fields and in open woods in the midwestern United States, and
may play an important role in succession. The species is made up of
apomictic and sexual populations. It is usually absent from fertile sites and
from deep shade, but where present persists for a long time in large
numbers. Usually asexual populations of this species are more common in
mid-successional oldfields, and sexual populations, both male and female,
are usually found in open woodlands. However, in some fields, sexuals and
asexuals can be found together. Because *Antennaria* is a clonal perennial
plant, it can be cloned easily for experimental work. Multiple copies of the
same individuals can be made to study their response along gradients. We
compared the response of asexual and sexual plants along gradients of light
and nutrients, which are critical to *Antennaria* distribution (Michaels and
Bazzaz 1986, 1989; Fig. 8.15). We found that:

1. At all resource levels, sexual populations produced more total biomass
   but less reproductive biomass than asexual populations.
2. The allocation to reproduction by asexuals increased with an increase in
   resource level, whereas it was comparatively stable in sexual plants.

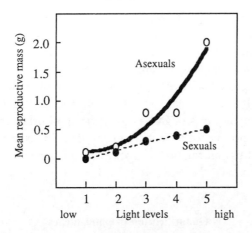

Fig. 8.15. Mean reproductive biomass of asexual and sexual genotypes of the mid-successional *Antennaria parlinii* in relation to light (modified from Michaels and Bazzaz 1989).

3. Within populations, the response of individual sexual genotypes was quite variable, whereas in asexuals, it was relatively homogeneous and broad.
4. Populations of sexuals had a greater between-genotype component of response breadth than did asexuals (mean = 8.75 vs. 3.33).

Thus, asexuals are well suited for the colonization of disturbed sites, as they are highly plastic in their growth on a wide range of resource gradients. They seem to possess broad genotypes and are opportunistic in reproduction. In contrast, sexuals are more adapted to competitive, late-successional habitats. They place an emphasis on vigorous vegetation growth, show a broad response in reproductive characters, and produce a diverse range of individuals along resource gradients.

While the response of plants to individual resource gradients in their habitat can tell much about their ecology in nature, many environmental factors simultaneously impinge on the plant proximally, modifying its response and ultimately its evolution. Therefore, while knowing the response of plants on single environmental gradients is useful, data on its response on all the critical gradients of its environment are needed to understand its behavior and evolution. Experiments that simultaneously consider several resource gradients that are critical to plant function are necessary for a clear understanding of niche relations in plants. Because the shape of the response of a plant on one environmental gradient can be greatly influenced by the levels of other critical resources, an assessment of a plant's response to its natural environment requires a multifactored approach. Unfortunately, multifactor responses of plants have not been adequately examined. Furthermore, in the few studies using two gradients, the response was found to be unpredictable from the responses on the gradients when they were used singularly (Parrish and Bazzaz 1979, Zangerl and Bazzaz 1983*b*). The experimental design of multifactor experiments may not be straightforward. The appropriate ranges for each resource may not be easy to determine prior to experimental work. Nevertheless, this information is critical to an understanding of plant resource-response.

Quantification of plant niches

There are several methods for assessing the results of these experiments (Smith 1982, Petraitis 1981), and for analyzing response data on single gradients, e.g. Levins' B, Colwell and Futuyma's 1971 method of calculating weighted niche breadths, partitioning of niche breadth into between- and within-phenotype (Roughgarden 1979), and modification of the Finlay and Wilkinson method for magnitude of response (Garbutt and Zangerl 1983). However, multivariate methods (e.g. Carnes and Slade 1982, Dueser and Shugart 1982, Van Horne and Ford 1982) have also been used.

Borrowing from mathematics, it is now possible to quantify niche volume more accurately (see Carlton 1993 for further detail). In order to quantitatively describe whole plant growth response surfaces, we used descriptive models that combine techniques from multiple regression and vector calculus. Vector models have broad application in the physical sciences, especially in fluid dynamics and electromagnetic theory. These models allow us to estimate growth at different combinations of resources, and predict, for any given combination of resources, the growth response to additional inputs of different resources. In vector terminology, a response function generated by multiple regression techniques is mathematically equivalent to a scalar field (Marsden and Tromba 1981). For application to niche analyses, abiotic resources are the independent variables in the response function, and some measure of plant growth is the dependent variable. A two-dimensional scalar field can be represented graphically as a series of curves, each of which defines all points at which the response function assumes a particular value. Similarly, a three-dimensional scalar field can be depicted as a series of level surfaces.

Associated with each point in the scalar field is a vector known as the gradient. The gradient vector for a three-dimensional scalar field is defined as:

$$grad\phi = \left(\frac{\delta\phi}{\delta x}, \frac{\delta\phi}{\delta y}, \frac{\delta\phi}{\delta z}\right) = V_x, V_y, V_z$$

where V_x, V_y, and V_z are the components of the gradient vector. The magnitude of a three-dimensional gradient vector is calculated as:

$$|grad\phi| = \sqrt{(V_x^2 + V_y^2 + V_z^2)}$$

The set of all gradient vectors throughout the scalar field is a vector field. The change in magnitude of a vector field within a small area (or volume in higher dimensions) is measured by a scalar quantity called the divergence. The divergence of a three-dimensional vector field is calculated as:

$$div \cdot \mathbf{V} = \frac{\delta V_x}{\delta x} + \frac{\delta V_y}{\delta x} + \frac{\delta V_z}{\delta x}.$$

Divergence measures the curvature of the growth response function. Negative divergence, for instance, indicates decreasing growth response to resource inputs at higher resource levels. Gradient vectors and vector fields have some interesting and useful properties with regard to quantitatively describing plant responses to multiple resources. All gradient vectors are perpendicular to the level curve or surface from which they originate, and they point in the direction of greatest change in the response function. The magnitude of the gradient vector is equal to the change in the scalar field associated with a unit change in each independent variable, in this case the resource. Each vector component estimates the change in the response function associated with unit change in the corresponding independent variable. In our models, vector components indicate the relative contribution of each resource to the magnitude of the growth response.

We have used these techniques to study the response of birch seedlings to resources on microsites created by an experimental simulated hurricane blow-down at the Harvard Forest (Carlton 1993). Vector models revealed that white birch seedlings responded fundamentally differently to light and nitrogen. In low light conditions, small increases in photon flux density (PFD) greatly enhanced diameter growth rates, but seedlings in high light were relatively unresponsive to increased light levels. Seedling response to increased nitrogen availability, conversely, was greatest at high nitrogen levels. The magnitude of the growth response to additional light was positively related to nitrogen availability, whereas the response to added nitrogen was similar at all light levels. The decreasing benefit of small increases in light at higher light levels was quantified as negative divergence. The vector models outlined above thus appear to have considerable potential as descriptive modeling tools. They offer both analytical precision and descriptive geometrical appeal, being of particular use in providing quantitative descriptions of such concepts as multiple resource limitation, congruence, and compensation.

It seems that there is no reason to abandon the niche concept in the analysis of plant communities, despite its obvious limitations. Niche theory will continue to enrich our understanding of the organization of plant community along successional gradients. However, in order to satisfactorily determine multigradient responses, and more convincingly determine plant niches, the following points must be considered:

1. The relevant resources and controllers of the plant's niche must be correctly identified. Light, water, nutrients, CO_2, and temperature are important to all plants and must be included.
2. The range of potential patches a species can occupy in nature and the quantification of the levels of environmental factors in these patches should be obtained.
3. The importance of biological resources may be different for different plant species. Various degrees of specialization in pollination, dispersal, herbivory, mycorrhizal associations, and pathogens are being discovered in many ecosystems.
4. Little is known about factor interaction and the degree to which plants can compensate for excesses or shortages of one resource through acquisition of other resources. For example, how much reduction in light could a plant tolerate if it were to receive an additional quantity of phosphorus?

9

Ecological and genetic variation in early-successional plants

Morphometric analysis, isozymes, and DNA fragment analysis indicate the presence of much genetic variation in many plant populations even for fitness-related characters (e.g. Solbrig 1981, Ennos 1983, Antonovics 1984, Mitchell-Olds 1992). The maintenance of this variation in natural populations, even for characters that are assumed to be subject to strong selection, is a critical area of research in plant population biology. Although there are some general trends relating the level of genetic variation to habitat preferences in plants, it has long been recognized that the relationship between genetic variation and the performance of populations in their natural habitats can be rather complex (e.g. Ayala 1982, Schaal 1985). In many situations, genetic variation does not necessarily translate into morphological and physiological variation directly relevant to the performance of plants. The general relationship between the level of variation in populations and their habitat preferences has been analyzed in some plant species (e.g. Hamrick *et al.* 1979, Loveless and Hamrick 1984), but few studies have closely examined the patterns of genetic variation in plant populations along successional or disturbance gradients. However, because of the high level of inbreeding among early-successional annuals, we expect their populations to have a lower number of genotypes relative to plants of late succession that are usually outcrossers (Parrish and Bazzaz 1979). Therefore, in early-successional plants, variation in the population is expected to be found within, rather than between, individual genotypes; i.e. individuals have much flexibility in response. That is to say, the large niches of these plants are the result of the presence of a limited number of similarly large niched genotypes, but not the result of the presence of specialized genotypes with a broad collective response (see Chapter 8).

It has been clearly demonstrated in many experimental populations that selective pressures can change gene frequencies and enable some members of the popula'ion to adapt to a changing environment. For example, Bradshaw and Hardwick (1989) have shown that natural populations that are subjected to strong selection by the presence of heavy metals in their environment become tolerant of these heavy metals. Long-term fertilizer application selects for specific and appropriate genotypes (as has been shown by Snaydon and Davies 1972). It has been suggested that strong biotic pressures (e.g. by neighbors) can also result in selection in *Trifolium repens* (Turkington and Harper 1979) and other pasture plants (Turkington 1983, Aarssen and Turkington 1985). Nevertheless, many natural plant populations still contain much genetic variation.

Despite the great strides made in ecological genetics in plants, there has not been a clear differentiation between 'genetic variation' and 'ecological variation,' nor have the implications of genetic and phenotypic variation to the niche size and habitat range of the population been made clear. *Genetic variation* refers to DNA-based differences among individuals, and is usually assessed by measuring morphometric characters, electrophoretic enzyme analysis, DNA restriction fragments analysis, DNA sequencing or other techniques of molecular genetics. Therefore, *phenotypic variation*, upon which selection operates, is the product of genotype–environment interactions. Phenotypic variation, which may not be entirely genetic, can have evolutionary consequences by affecting relative fecundity and the quality of the offspring produced (see Sultan 1987). In some situations, especially when the plant is highly flexible, the environment may induce changes in the phenotype that are far more critical to performance than the actual identity of the genotype. For example, growth and reproduction in plants can be greatly influenced by the environment (Sarukhán *et al.* 1984). Thus, in some situations similar phenotypes, irrespective of their genotypic origin, may occupy a similar range of habitat patches in the field or a given range on an environmental gradient. Therefore, phenotypic variation can be intimately related to 'response breadth' on a single gradient and similar to 'niche space' on several gradients.

Unlike genetic variation, *ecological variation* describes the range of habitats the plant can occupy (Bazzaz and Sultan 1987). One population, or even an individual genotype, can persist in a relatively narrow range of habitats (or breadth of environmental gradients), while another population or genotype can persist in a relatively wide range of habitats. In such a case, the former would have little and the latter much 'ecological variation.' The habitat range occupied by a population is best assessed experimentally in a comparative way, as it would be difficult to scale abstractly the range of habitats occupied by various populations, particularly when more than one habitat factor is considered. As mentioned earlier (Chapter 7), the diversity of habitat patches or the range occupied by an individual or a population on a gradient can be greatly modified by the presence of neighbors and their identity. Thus, the 'ecological variation' of a population in the field reflects the degree of gene expression, the phenotypic variation resulting from specific genotype–environment interactions, the identity of neighbors, and interactions among environmental variables of the habitat that impinge on the population. In the presence of competitors, ecological variation in the field is thus related to the Hutchinsonian notion of the realized niche.

In order to explore the relationship between positions along successional gradients and the level of variation in populations, we ask the following:

1. How is the level of genetic variation found in successional plant populations maintained in nature despite the presence of selective forces?
2. Are populations of successional plants made up of equally broad niched individuals or are they combinations of some narrow and some broad individuals?
3. Does the relative mix of genotypes with narrow vs. wide niche breadth in a population change along successional gradients?
4. Is there a relationship between successional position and the degree of phenotypic variation in populations?
5. Does ecological variation (niche volume) correlate with genetic variation?
6. Is the relationship between genetic variation, phenotypic variation, and ecological variation necessarily positive?

We identify several factors that can promote variability in successional plant populations. These include environmental heterogeneity, seed bank longevity, response flexibility, and clonal integration.

Environmental heterogeneity

Environmental heterogeneity occurs at several scales (see Kolasa and Pickett 1991), even at the scale of seedlings (e.g. Harper 1977, Blom 1978). Because of small-scale environmental heterogeneity, selection for variability in the initial life stages of individuals may be critical to the structure of populations and the maintenance of variation. Under some conditions, this small-scale patchiness is so strong that it can actually override genetic differences among seedlings. For example, using the annual plant *Abutilon theophrasti*, Hartgerink and Bazzaz (1984) created small (seedling-size) patches of high nutrients, low nutrients, small stones above the soil surface to impede seedling growth, depression in the soil to simulate footprints, etc.

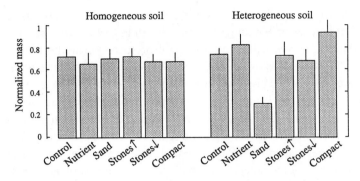

Fig. 9.1. Despite uniform seedling emergence time, seedlings of *Abutilon* accumulated biomass at different rates in a heterogeneous soil versus corresponding positions in a homogeneous soil (modified from Hartgerink and Bazzaz 1984).

These patch types were chosen to represent factors that could be unpredictable in both space and time in the environment of seedlings in early-successional fields. Seeds collected from a large area were individually sown in these patches at random. A uniform environment corresponding to the same patch locations was also established and sown with seeds as well. At maturity, the plants in the heterogeneous patches differed significantly from each other in several growth and fecundity parameters, while those in the homogeneous soil were much more uniform in these performance characteristics (Fig. 9.1). Thus, we demonstrated that the position of an individual in the population hierarchy, irrespective of its genotype, can be greatly influenced by small-scale stochastic events that can be both spatially and temporally unpredictable. Therefore, in the absence of strong and consistent selective pressures, there would be little chance for microevolution of different genotypes. Genotypes that may be selected against in some homogeneous environment would be retained within the population through the action of heterogeneity, resulting in the maintenance of variation in these populations (e.g. Antonovics 1971, Spieth 1979, Hedrick 1986). We therefore assume that at least some of the variation in successional plant populations is related to the magnitude of environmental variability in their habitats.

In plant populations asymmetrical competition is common. In this case larger individuals take more of the environmental resources relative to their size (Weiner 1988, 1990). Strong competitive asymmetry can by definition lead to exclusion of smaller individuals or drastically reduce their contribution to the seed bank. When size differences result from genetic differences among plants, competition may lead to an overall selective reduction in genetic variation in the population. However, if the differences among neighbors are due to chance (environmental stochasticity), then dominance asymmetry may lead to the death of the smaller individuals but not a long-term reduction in the standing genetic variation of the population. This is particularly strong when seeds are long-lived in the seed bank.

Environmental variability has been shown to promote species diversity in plant communities (Bazzaz 1975, Whittaker 1975, Grubb 1977, Huston 1994). Similarly, at the population level, theoretical analyses (e.g. Ennos 1983, Rice and Jain 1985) have shown that environmental heterogeneity contributes to the maintenance of genetic variability in their populations. The spatial and temporal heterogeneity characteristic of the plant's changing environment imposes complex and fluctuating selective pressures. The changing relative fitness of various genotypes in a population, caused by variability in site condition, recruitment patterns and seedling emergence time, can mean that the same genotypes may be relatively less fecund in one year and relatively more fecund in another year or another location. Complex selection pressures may fail to produce adaptations to the local environment because it is in a state of flux. Different patches may have different levels of light, moisture, nutrients, herbivore and pathogen loads, etc. All these may promote variability in populations, particularly if the population occupies several of these patches. Moreover, spatial heterogeneity can be perceived by an individual at many scales. As the individual plant grows in size from a small seedling to a mature individual, it samples more of the habitat and may therefore experience a wide range of variability throughout its life. We studied field populations of the annual *Phlox drummondii*, which occurs in habitats that are repeatedly disturbed and thus continually early successional. Natural populations contain both red-flowered and white-flowered individuals, and the two flowers, color morphs are maintained in natural populations for many generations (Levin 1984). Plants with white flowers are shown to be inferior to those with red flowers in controlled environments. They are greatly reduced in frequency in crowded situations (Bazzaz *et al.* 1982). Thus, we propose that the white flowered morphs are maintained in natural populations by environmental variability created by patchy disturbance.

Persistence of less fit genotypes has also been shown to occur in several other species, such as *Solanum delcamara* (Clough *et al.* 1980). Because fecundity, especially of early- and mid-successional herbaceous plants, is highly influenced by relative performance in changing competitive neighborhoods, even unstable polymorphism can be maintained for some time (Sultan 1987). Populations of *Phlox drummondii* found in patches differing greatly in soil moisture, nutrient content, and temperature regime do not differ in response to controlled gradients that correspond closely to the factors in their particular habitats (Schwaegerle and Bazzaz 1987). We assume that the environment within each of these habitats changes from year to year, selecting for broad genotypes and promoting phenotypic variability.

In order to understand how variation is maintained in early-successional plants, we must first demonstrate that strong and consistent selection does lead to change in the genetic structure of their populations, and that at least some of that change has adaptive value in the environment of selection and is not simply the result of drift. To test this, we used species of the early-successional annual community in a selection experiment for four generations on moisture, density, and diversity gradients, common selective pressures in the habitats of these species (Zangerl and Bazzaz 1984*b*). *Amaranthus retroflexus* and *Abutilon theophrasti* seeds were collected from several fields in a wide geographic area to ensure a high level of genetic variation in the populations. Random samples of the population of each species were subjected to selection for several generations on the three gradients. The soil moisture gradient ranged from permanently saturated to dry, and plants were selected from three levels of soil moisture: high (50%), medium (30%), and low (5%) percentage of dry weight. There were also two conspecific densities and three levels of neighborhood diversity. The experiments were run for four generations in two years, with the generations initiated in May of the first year and February, May, and November of the second year. For each gradient, from each generation, seeds of a selected line were sown along the entire gradient for the following generation. In order to find out what changes occurred during the selection process, individuals from each line and each generation were compared with individuals from the original population, which were all grown under uniform environmental conditions. A total of 30 morphological and physiological characters were measured on each individual. A principal component analysis (PCA) was performed to quantify the degree of divergence under selection. In both species, lines did diverge from the original population. Out of the 30 characters, 17 in *Amaranthus* and 20 in *Abutilon* significantly differed from the original population. These included photosynthetic rates, root/shoot ratios, total seed production, and reproductive ratios (reproductive biomass/vegetative biomass), all of which are critical to growth and fecundity. Lines that differed from the original population were also lower in overall variation, and in no case was a selected line more variable than the original population.

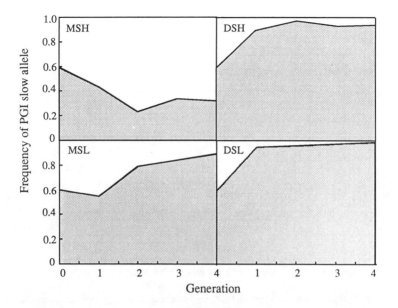

Fig. 9.2. Changes in phosphoglucoisomerase (PGI) allele frequencies during four generations of selection on moisture and diversity gradients (modified from Zangerl and Bazzaz 1984*a*). MSH, high moisture selection regime; MSL, low moisture selection regime; DSH, high density selection regime; DSL, low density selection regime.

With strong and consistent selection, alleles are quickly fixed in populations. Electrophoretic analysis of two phosphoglucoisomerase (PGI) homozygous genotypes showed that fixation of their frequency in *Amaranthus* occurred rapidly in the moisture selection regime but not in the diversity selection regime (Zangerl and Bazzaz 1984*c*) (Fig. 9.2). The relative frequency of each allele was different in the moisture-selected lines, with the slow allele (SS) more prevalent in the 'low moisture' line and the fast allele (FF) more prevalent in the 'high moisture' line. In the diversity selection regime, in contrast, there was no difference in frequency. What, then, might cause the differences in allelic frequency between the moisture selection regime and the neighborhood diversity selection regime? The moisture selection regime was consistent throughout the four generations because it was experimentally controlled. The diversity regime was not constant, although the same neighbors were used in all generations and they came from the same seed pool. Since the four generations were grown in different months and in different years in a temperate climate, the contribution of each species in the mixture varied. Thus, there was variability in the biological environment

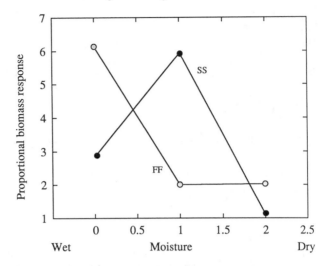

Fig. 9.3. Performance of plants with the fast (FF) and slow (SS) PGI-marked alleles on a soil moisture gradient. (FF) genotypes were selected on the wet end of a moisture gradient and (SS) on the dry end of the gradient (modified from Zangerl and Bazzaz 1984*d*).

from generation to generation. For example, *Setaria faberi* contributed greatly to total biomass in some generations and little in the others. Relative to its neighbors, *Setaria* germinates (Bazzaz 1984*a*) and photosynthesizes optimally at high temperatures (Wieland and Bazzaz 1975). Its contribution in various generations reflects its physiology, in that *Setaria* was more important in assemblages started during the warm period of the experiment. Similarly, the density treatment created a complex, changing environment rather than a consistent, simple gradient. In contrast, the consistent and strong selection on the moisture gradient produced genotypes adapted to the appropriate soil moisture. The two PGI-marked genotypes show this very clearly. When these genotypes were grown on a moisture gradient, homozygous slow (SS) genotypes, which became predominant in the low-moisture selection regime, performed better in low moisture conditions. In contrast, the homozygous fast (FF) genotypes performed better in the high moisture conditions. Biomass production was highest on the wet portion of the gradient for the FF genotype and on the moderate moisture level for the SS genotype (Fig. 9.3). Furthermore, seed germination was higher and occurred more quickly in the FF genotype in low oxygen concentrations associated with high moisture conditions. The reverse was true for the SS genotype (Fig. 9.4).

Three of four natural populations sampled in different fields within an

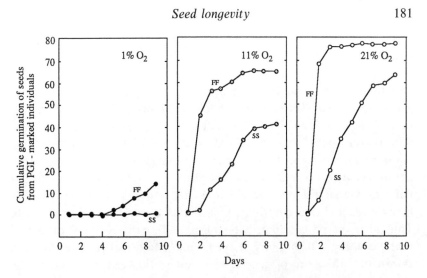

Fig. 9.4. Differences in seed germination in response to O_2 levels of the fast (FF) and slow (SS) PGI marked individuals (modified from Zangerl and Bazzaz 1984*d*).

area with a radius of about 8 km had both alleles but in different ratios. Therefore, this study demonstrated that (a) consistent selection on simple environmental gradients can lead to a reduction of variation and to lines that are adapted to perform well in the appropriate environment, and (b) the changing biological neighborhood and the varied physical environment associated with differential resource use can maintain variation in populations. Interestingly, despite the decline in overall variation with selection in several morphological, physiological, and electrophoretic characters, there was limited divergence in ecological variation (niche space) among the lines. This result contradicts the niche width variation hypothesis of Van Valen (1965) that predicts that genetic variation and niche breadth are positively correlated.

Seed longevity in the seed bank

Seeds of many early-successional plants have various dormancy mechanisms, and their germination may be induced by disturbance (see Baskin and Baskin 1988 for a review). Many seeds remain viable in the seed bank for many years (Fenner 1992, Thompson 1992). The length of persistence in the seed bank can differ among species of the same community (Bakker 1989). Because of longevity and environmental variation, the seed bank of early-successional habitats is likely to be a mixture of many genotypes and species that have accumulated over several generations. In fact, the seed bank constitutes a large 'genetic pool,' which can be much more variable than what is recruited through germination in any given season (see Parker *et al.* 1989). This hidden variation can be much larger than 'apparent variation.' Because of the extended seed longevity in many of these early-successional species, it is probable that different years with different rainfall patterns and different levels of soil moisture and soil temperature during recruitment can bring out from the seed pool more of the 'appropriate' genotypes for that year. Different genotypes in the seed bank are expected to do well in different years with different environmental patterns. Genotypes that grow particularly well in a given year produce many seeds and thus contribute greatly to the seed bank (Bazzaz 1979). Thus, all genotypes can contribute substantially to the seed pool in different years, creating a highly diverse seed bank. Mathematical modeling has shown that the store of genetic variation in the seed bank allows the population to resist selection in response to short-term environmental fluctuations (Templeton and Levin 1979), and can also prevent loss through genetic drift (Loveless and Hamrick 1984). It has also been shown mathematically (Brown and Venable 1986) that when different kinds of seasons occur unpredictably, evolutionary specialization for seedling response will be opposed, and high germination caused by occasional disturbances will increase the variance in fitness and thereby oppose selection. In contrast to this, seeds of late-successional plants tend to be short-lived (Bazzaz 1979). They are either eaten by predators (Crawley 1992), killed by pathogens (Roberts 1981), or quickly germinate and become part of the seedling bank (e.g. Whitmore 1984, Martínez-Ramos *et al.* 1988*b*, and Alvarez-Buylla and García Barrios 1991). Therefore, seed banks are expected to contribute relatively less to the maintenance of variation in late-successional populations.

Perennial plants of more stable ecosystems are not without mechanisms to maintain variation. The perennial habit is analogous, in terms of selection, to seed banks in annual plants, in that they are stores of variation that can become available to selection over several seasons (e.g. Jain 1979). The annuals accomplish this by having progeny stored in the seed bank, only a small part of which are expressed as aboveground plants in any given season. Long-lived perennial plants can integrate environmental heterogeneity over time. Because of their cumulative growth, their capabilities to store critical resources, and the fact that they can delay their reproduction to more favorable seasons (masting), perennial plants can 'buffer' much of within-season and between-season environmental variability and thus maintain variation. In this way, early- and late-successional plants differ

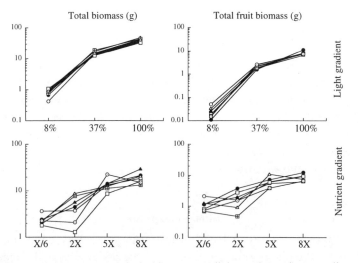

Fig. 9.5. Total biomass and fruit biomass to light and nutrient gradients of genotypes of *Polygonum persicaria* (modified from Sultan and Bazzaz 1993a,c).

only in the degree to, and possibly the mechanisms by which variation is maintained.

Individual response flexibility

Response flexibility, which is particularly high in early-successional plants, may be a major contributor to the maintenance of variation in plant populations. This flexibility, the inherent capacity of a genotype to undergo specific (adaptive) phenotypic modification in response to a specific environmental influence, is also referred to as the *norm of reaction* (Schmalhausen 1949). Our work on the early-successional annual *Polygonum* shows that genotypes have very broad norms of reaction on several environmental gradients (Sultan and Bazzaz 1993a,b,c; Fig. 9.5). Species of the early-successional community generally have broad norms of reaction on several environmental gradients, including moisture (Pickett and Bazzaz 1976, 1978a), light (Sultan and Bazzaz 1993a,b,c), nutrients (Parrish and Bazzaz 1982b,c), and use of pollinators (Parrish and Bazzaz 1979).

Species that respond with a high degree of flexibility can generate appropriate (adapted) phenotypes in different environments (Sultan 1987). In addition, different genotypes may produce similar phenotypes in the same environment (Levin 1988, Sultan and Bazzaz 1993a). Since selection acts on the phenotype, genotypes in this case are not differentially available to selection, so that the initial level of genetic variation may be maintained in the population. Despite the fact that early-successional plants have been shown to possess more flexibility in response than late-succession plants (e.g. Zangerl and Bazzaz 1983a,b), the former do not necessarily always have less genetic variation than the latter (Jain 1990). Many other events, such as seed dispersal, gene flow by pollen, and the degree of inbreeding, influence the genetic structure of a given population (see Levin 1984). Depending on the circumstances, these parameters can work to oppose or promote genetic variation in all populations.

Clonal integration

Clonal individuals are capable of expressing modules (leaves and ramets) of different morphologies and physiologies as the clones grow and extend into different environmental patches. While the rate and the direction of expansion can vary among clonal species and habitats, physiological integration, or the ability of parts of a clone to move resources and hormones across the entire clone, has now been shown to occur in many plants. Photosynthate translocation is especially well documented for both early- (Schmid and Bazzaz 1994) and late-succession clonal herbs (Ashmun *et al.* 1982). By clonal growth, individuals can integrate environmentally different patches, thus reducing the probability of ramet mortality in unfavorable patches (Bazzaz 1984b, 1991a).

In mid-successional habitats, *Solidago* plants can grow in a variety of neighborhoods, especially those including other *Solidago* species, *Aster*, and *Poa*. As a *Solidago* genet grows larger, it can come into simultaneous contact with patches dominated by different neighbors having different levels of resource uptake capacities, different resource depletion zones, and different degrees of resistance to invasion (see Chapter 3). Clonal integration is expected to allow the genet as a whole to do well, even when parts of it are found in less hospitable neighborhoods. We tested this hypothesis by growing *Solidago* plants as connected individuals and as severed ramets in three different competitive neighborhoods (Hartnett and Bazzaz 1985b). The ramets of the intact genet performed similarly in the different neighborhoods, while severed independent ramets performed differently in the different neighborhoods (Fig. 9.6). Clonal integration also reduced the negative effects of herbivory on ramets. Therefore, physiological integration of connected ramets can buffer genets against patch-specific, fine-scale selection. Another manifestation of integration is the equitable spread in all

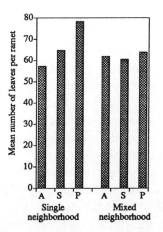

Fig. 9.6. Performance of *Solidago*, disconnected (left) and connected ramets in different competition neighborhoods. A, *Aster*; S, *Solidago*; P, *Poa* (modified from Hartnett and Bazzaz 1985*b*).

directions of many clonal species even when their ramets come into contact simultaneously with different neighbors. Therefore, genet diversity is maintained in the fields, as most of the genotypes that become established early in colonization remain. We observed no genet mortality over five years of monitoring 20 clones with distinct morphological and phenological characteristics. Population growth occurred largely through clonal spread and an increase in the number of ramets. Thus, selection forces do not determine survivorship of clonal plants, and the genetic diversity of populations is maintained.

Environmental influences on plant fitness

Fecundity and other fitness-related characters can be greatly influenced by the environmental circumstances not only under which plants grow but also under which they reproduce (Roach and Wulff 1987). For example, differences in seed size can be induced by changes in the level of resources in the environment (see Schaal 1984, Alexander and Wulff 1985), and may have great influences on the emergence, survival, growth, and reproductive success of the resulting progeny (Stanton 1984*b*, Antonovics and Schmitt 1986). Maternal environmental effects can last for as many as three generations (Alexander and Wulff 1985, Miao *et al.* 1991*b*), and maybe even more. These non-genetic effects on plant fitness may mask genotypic differences among individuals and thus oppose selection and promote the maintenance of variation (see Schmitt *et al.* 1992).

In *Ambrosia trifida*, an annual in early-successional fields, timing of germination and emergence is greatly influenced by the depth at which seed is buried in the soil. Early germinants, regardless of genetic identity, have a higher probability of preempting resources, occupying higher positions in the size hierarchy, and reproducing in disproportionately greater numbers, relative to late-emerging germinants. Therefore, recruitment time, depth of seed in soil, location in micropatches of differing resource availability, and site hospitality can influence the genetic structure of a population of both exposed living individuals and those hidden in the seed bank. In this species, competitively suppressed individuals function mainly as females (Abul-Fatih and Bazzaz 1979*b*). They develop no male flowers and have a much higher reproductive allocation relative to their taller neighbors. Furthermore, the taller individuals in a population suffer much more insect seed predation than their suppressed neighbors. Therefore, suppressed individuals can contribute disproportionately more seed to the population, irrespective of genetic identity, since suppression is largely determined by emergence time, which is mostly a function of depth of seed. These biotic influences can also operate in a frequency-dependent fashion and promote the maintenance of genetic variation in a similar way (Ayala and Campbell 1974). For example, less common genotypes in a population can do better than expected if their more frequently found neighbors are subject to herbivory by virtue of their higher apparency. In this case both less common and more common genotypes are maintained in the population. Existing genetic variation in successional populations will be the result of the interactions between forces of selection and forces that maintain variation.

Changes in the genetic structure of persistent populations during succession

Long-lived perennials that persist in successional fields, experiencing a changing environment and interacting with different neighbors, can express change in the phenotypic structure of their populations. Despite high levels of variation in successional plant populations, selection for certain genotypes with successional change has been proposed (Gray 1987). Differences in phenology and reproductive allocation were found among populations from fields of different successional ages (Roos and Quinn 1977). A decline in variance of certain characters with time has been observed (McNeilly and Roose 1984, Aarssen and Turkington 1985, Gray 1987, Taylor and Aarssen 1988). As succession proceeds, populations of most annual species are represented mainly in the seed bank and by a few reproducing individuals in small patches where soil disturbances occur. Individuals of a

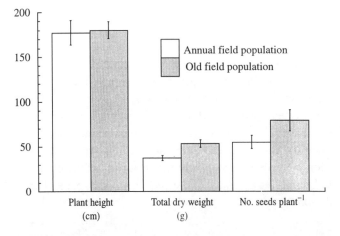

Fig. 9.7. Height, weight, and seed number of *Ambrosia trifida* populations from a 1-year field and from a persistent population in the field for 15 years (modified from Hartnett *et al.* 1987).

few species of annuals may persist as reproducing individuals during succession, and they may interact strongly with the late-succession perennials. What influences might this interaction have on the genetic structure of these persisting annual populations? Will there be a reduction in genotypic diversity with time in these persisting species?

The giant ragweed (*Ambrosia trifida*) is one such annual that can persist in some successional fields. Two populations of *Ambrosia*, one from an annually disturbed section of a field, and another from an adjacent section of the same field left undisturbed for 15 years, were compared for several characters. It is worthwhile to note that these two populations were part of one population prior to the initiation of a long-term successional study. Transplant experiments in the field showed several morphological and physiological differences between the two populations. Individuals from the 15-year field emerged earlier, had greater numbers of leaves, greater biomass and seed production, and higher reproductive ratios relative to plants from the young field (Fig. 9.7). Furthermore, individuals from the older field were competitively superior when grown with individuals from the younger field (Hartnett *et al.* 1987). Since they were initially members of one population, the differences between the two populations may have resulted from selection imposed by the changing neighborhood over successional time. But, although the older population showed less variability for half of the traits examined, there were no statistically significant differences between the two populations in overall variation.

Whether fragmentation of the habitat, caused by patchy occupation by different species of succession, leads to the fragmentation of persistent populations and the maintenance of overall genetic diversity has not been systematically investigated. Gray (1987) concludes that the decline in variation with successional age observed in some populations may simply be a consequence of a decline in numbers. Furthermore, in species that possess a persistent seed bank, it is likely that many genotypes remain in the seed bank and that these genotypes emerge once more only after another disturbance. In this case, the difference in genotypic diversity with the progression of succession is mostly the result of differences in the relative amount of genotypic diversity 'expressed' as growing plants and 'hidden' in the seed bank.

Is there a relationship between genetic variation and ecological variation?

The 'genetical theory of natural selection' assumes that selection acts on morphological variants that are determined directly by certain genes, rather than by the developmental properties of the entire organism (see critique in Sultan 1993). In this view a given gene produces a phenotypic trait, and therefore there is close correspondence between genetic and phenotypic variation. However, recent molecular data have revealed a general lack of correspondence between genetic variation and the phenotypically expressed variation directly relevant to selection (Jain 1990). As we have seen, developmental events can be greatly influenced by the environment.

As discussed above, the relationship between environmental variability and the variability of response of plant populations (ecological variability) has been of much interest to ecologists and population geneticists. A related and critical question that has been less frequently considered is whether there is a positive relationship between genetic variation and ecological variation (niche space). Are broad-niched species more genetically variable than narrow-niched species? This notion has been embodied most formally in the niche-width variation hypothesis articulated by Van Valen (1965). Since early-successional plants have a higher level of ecological variation (niche space) (Chapter 8), we then ask: do early-successional plants have a higher degree of genetic variation than do late-successional plants? Or are there other mechanisms by which early-successional plants can be large-niched (possess high levels of ecological variation) and still have low genetic diversity? To answer these questions, we should consider the structure of the population response.

The structure of the population niche and the proportion of specialists and generalists in a population are relevant to the response of the population to selection. A population made up of 'generalists' loses no genotypes under selection pressure. Its genotypic diversity does not change, despite a reduction in its ecological diversity (niche volume). In contrast, a population made up of specialized genotypes would lose some genotypes under specific selective pressures. In such cases, there will be a strong correlation between a population's genotypic diversity and its niche volume. Even in the case of 'specialist' genotypes that perform poorly in certain conditions, genetic diversity may be maintained in populations in which conditions vary due to the extreme heterogeneity of soil moisture and other aspects of the plant environment in early-successional sites (Zangerl and Bazzaz 1983*a,b*, Sultan and Bazzaz 1993*b*, Thomas and Bazzaz 1993). Spatial and temporal environmental variability are thus major influences on standing genotypic, allelic, and phenotypic variation in successional populations (Zangerl *et al.* 1977).

Variation and the structure of the population response

Several investigators (e.g. Lewontin 1957, Levins 1968, Roughgarden 1979) have suggested that broad response to the environment by a population could be achieved in two ways: either the population has a high level of genetic variation such that each variant occupies a portion of the habitat, or each individual of the population has a broad response such that it can occupy many habitat patches. In the former case, there will be a positive relationship between genetic variation and ecological breadth, whereas in the latter, a few individual genotypes can cover the entire range of environmental patches that are likely to be encountered by a species. What, then, is the genetic structure of a population in early-successional plants?

Despite compelling evidence for the evolution of plasticity among plants (Bradshaw 1965, Sultan 1987) and the preponderance of highly plastic genotypes in populations of early-successional plants (Bazzaz 1979), these populations can contain some narrowly adapted (specialized) genotypes as well. In natural populations, there is likely to be a combination within any population of 'broad-niched' and 'narrow-niched' genotypes. The relative number of 'generalists' and 'specialists' in a population depends on several factors, including the breeding system, dispersal patterns, and the scale and amplitude of environmental heterogeneity. By breaking the genets into ramets (e.g. tillers in grasses), the performance of *individual* genotypes in a range of environments can be examined (Clausen *et al.* 1948, Antonovics *et al.* 1987). In species that cannot be cloned vegetatively, it is often possible to use axillary bud enhancement (which keeps within-genotype clonal variations to a minimum) to examine the response of the individual genotype breadth of annuals on various gradients. We have studied the structure of the population niche of the early-successional annuals *Abutilon theophrasti*, *Polygonum pensylvanicum*, and *Polygonum persicaria*.

Data on the response of *Polygonum persicaria* genotypes from a repeatedly disturbed site (Sultan and Bazzaz 1993*a,b,c*) demonstrate that there is a very high level of plasticity in genotype characters relating to resource capture and use. This plasticity is associated with the ability of those genotypes to tolerate an extremely broad range of environments. Under severe nutrient-limiting conditions, these genotypes shared not only plastic response, such as increased root/shoot ratio, but surprising constancy in such functionally essential characters as leaf area ratio, leaf nitrogen content, and propagule nitrogen content. Furthermore, different genotypes converged plastically on functionally adaptive responses to light, moisture, and nutrient conditions. Such high degrees of plasticity exist even in habitats where the range of environmental resources commonly encountered is less than what any genotype could accommodate. If plasticity entails little or no cost, the evolution of plastic generalists rather than specialists may be favored (Sultan 1993).

10

Coping with a variable environment: habitat selection and response flexibility

Habitat selection

Except for a few extremely cosmopolitan, weedy species that can be found in an extremely wide range of environments, plant species generally occupy a subset of possible habitats to which they have adapted in terms of growth and reproduction. Failure to disperse in all suitable patches, competition from other plants, and other biotic factors limit plant distribution. Many plant species have evolved attributes that allow them to function in a relatively wide range of habitats, despite their relative inability to change locations (although plants can and do forage for resources). Under the influence of environmental variation in their habitats, plants have evolved several mechanisms to cope with this variability. These mechanisms of adaption to the environment include choice of the specific 'appropriate' habitat, and expanded response flexibility.

It is expected that as the environment of a patch changes, species can move to a more favorable patch; i.e. plants will have habitat choice. In fact, early-successional plants disperse to young patches as their current location is modified to become more suitable for later-successional species. The highly developed theory of habitat selection and its intimate relation with foraging, species packing and coexistence is based largely on studies of mobile animals (Rosenzweig 1991). Therefore, much of it is not directly applicable to plants (see Bazzaz 1991*a*). Currently there is no equally developed habitat selection theory for plants. For all plants, there are 'choice habitats' that share the following attributes:

1. Supply required resources (e.g. light, water, nutrients, etc.) in sufficient and balanced quantities for successful growth and reproduction.
2. Provide required mates, pollinators, dispersers, and symbionts.
3. Are relatively free of herbivores, predators and pathogens, except those that attack their competitors.

Despite their general immobility, plants can choose their habitats. Seed dispersal is particularly important for this process. Plants have evolved several means to detect and reach their 'choice habitats.'

1. *Relatively wide propagule dispersal.* Many plants of early-successional habitats produce numerous small highly dispensable seeds, which are capable of arriving at practically all patches including those that are suitable for their growth and reproduction. This is the case for so-called 'r-selected' species, especially those with the appropriate seed morphology for dispersal by wind. Mechanisms for habitat choice also include dormancy and other highly evolved physiological controls on germination. The early-successional plant species that do not efficiently disperse in space usually disperse in time; their seeds are long-lived and are recruited from the seed bank when the environmental conditions in the patch become suitable again.

2. *Targeted dispersal to favorable habitats.* Seed dispersal is usually thought of as a very passive phenomenon on the part of plants, but, like pollination strategies, some plants deliberately attract loyal dispersers that will move their seeds to appropriate habitats. Plants' propagules can have morphological characteristics that act as rewards to dispersers; these may be effective means for arrival in preferred habitats (Beattie and Culver 1979, Herrera and Jordano 1981, Willson 1983, Hanzawa *et al.* 1988). These preferred sites are usually located away from the parents, which may harbor pathogens and seed predators. Eliasomes, arils, fleshy fruits, and explosive dehiscence can facilitate transport of seeds far from the shade and pests associated with their parent plant (e.g. Augspurger 1983*a*,*b*, 1984). Animal ingestion of seeds may have an added value in that it can break their dormancy, usually by scarifying the seed coat (Fenner 1985).

3. *Dispersal with a supply of some required resource.* In some species dispersal by animals, e.g. large mammals such as cows or elephants may also enhance germination of seeds in nitrogen-rich dung, which can significantly increase seedling growth (Alexandré 1982, Dinerstein and Wemmer 1988). Dung may also supply moisture which is imbibed by the seed, causing the initiation of the germination process. In cases of dispersal in dung, the spatial pattern of the plant community can be greatly influenced by the patterns of movement of the dispersers and by their habitat preferences.

4. *Locations of foraging organs in habitat patches where resources are less contested by neighbors.* Plant modules can grow most actively in areas of

high resources (Harper 1977). The shape of an individual may therefore be highly influenced by its neighbors (e.g. Jones and Harper 1987, McConnaughay and Bazzaz 1991, Tremmel and Bazzaz 1993, 1995, Küppers 1994). Terborgh (1985) suggests that the position of the understory canopy of *Cornus florida* relative to the overstory maximizes light interception by *Cornus*. Moreover, many plants can concentrate new root and rhizome production in resource-rich areas of the soil profile, sometimes away from neighbors. Location of roots in different positions along the soil profile, which has been observed in many habitats, may also be a form of habitat choice (Salzman 1985, Salzman and Parker 1985). The integration of several patch types in order that more resources are available to the whole plants can also be achieved in some plants by clonal integration (Chapter 2). The actual movement to more favorable habitats can also be achieved by unequal clonal expansion. Some clones produce more ramets in areas of more favorable resources; as older ramets die, plants effectively move to a more favorable habitat. Preference to move closer to specific neighbors (or away from others) by clonal expansion has also been proposed (Turkington and Harper 1979, Aarssen and Turkington 1985, Schmid and Harper 1985, Silvertown 1987). All these plant behaviors are aspects of habitat choice by plants.

5. *Changes in life history.* Phenological shifts, the timing of life history events, may occur in some plants so that life history events coincide with the availability of resources such as water and warmth in seasonal environments, or with the peak of the activity of pollinators and the arrival of dispersers. For example, several herbs in temperate deciduous forests complete their life cycle early in the spring, when light, nutrients, and water from snowmelt are still available before the tree canopy closes. The introduced cool season grass *Poa pratensis*, which is most active in the grasslands of the American midwest and parts of the great plains, accomplishes most of its growth in spring and early summer and again in early fall when the associated C_4 grasses are less active.

Flexibility of response

As discussed in Chapter 8, terrestrial plants, which are both modular and sessile, cope with much of the variability in their environment by flexibility of response. Flexibility involves any or all of these actions: environmental *tracking*, *acclimation*, and *plasticity*.

Tracking environmental change

Environmental tracking means that plants may make short-term, reversible adjustments to rapid environmental changes, such as changes in stomatal conductance in response to rapid changes in vapor pressure deficit or light intensity. The rapid activation of the photosynthetic enzyme *Rubisco* with rapid increase in photon flux density, and its deactivation when plants are in low light, is another example of environmental tracking. However, some biochemical and especially morphological and anatomical changes can occur in a plant or a module only after a long exposure to a set of new conditions. These changes are either irreversible or are reversible only on an equally long time scale. Changes in *Rubisco* concentration, chlorophyll levels, or specific leaf area following a change in the immediate light environment are usually not reversible in the short term. When a light gap opens in the canopy and exposes individuals in the understory to high light, there may be a reallocation of nitrogen from light harvesting (chlorophyll) to CO_2 fixation activity (*Rubisco*). There is, of course, a continuous spectrum between quickly reversible responses and permanent, irreversible responses, as different plant structures and activities can change along different time scales. Thus, phenotypic response to the changing environment is expressed at a range of time and organizational scales within a plant. The contribution of phenotypic plasticity to the performance must be considered in relation to heterogeneity of the environment experienced by the individual plant as well as among individuals in a population.

Temporal variation in the environment, including oscillations between high and low light levels, can be associated with the diurnal and annual cycles in solar angle, day length, cloud cover, and the dynamics of the vegetation itself. It is expected that the time scale at which various traits of the plant change can differ (Table 10.1). For example, Ackerly and Bazzaz (1995*a*) found in an early-successional species of the rain forest of Mexico (*Heliocarpus appendiculatus*), that projected leaf area and specific leaf area respond on a 6-to-10-day cycle of oscillating high and low light, leaf area ratio on a 15-day cycle, and individual leaf area on a 30-day cycle (Fig. 10.1).

Fast tracking of changes in the environment is seen as a common strategy of early-successional plants. Because these pioneer plants are usually relatively short-lived, they must grow and reach reproductive maturity quickly; quick responses are an integral part of their life history. Opportunism, the taking up of resources as soon as they become available and before they subside or are taken up by other species, is a facet of this behavior. Such opportunism is likely to be selected for in environments where patterns of

Table 10.1. *Differences among traits of the early successional tree* Heliocarpus *in response to the frequency of switching between low light and high light (or vice versa). Traits differ in their response time. P2 represents switching on a 2 day cycle, P6 on a 6 day cycle, etc. Stars indicate levels of statistical significance*

	Cycle length				
Trait	P2	P6	P10	P15	P30
Specific leaf area (SLA)	*	**	***	***	***
Leaf area ratio (LAR)				***	***
Total leaf area				*	***
Root weight ratio					***

Source: From Ackerly and Bazzaz unpublished.

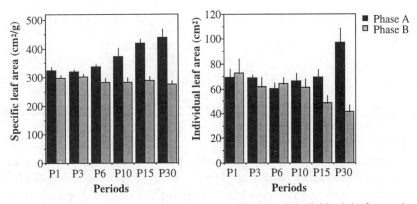

Fig. 10.1. Differences in specific leaf area (SLA) and individual leaf area in *Heliocarpus*, an early-successional tropical tree in high light–low light oscillating cycles of different speeds. SLA begins to diverge at a short time scale while individual leaf area diverges on a longer time scale (modified from Ackerly and Bazzaz 1995*a*).

resource availabilities are particularly unpredictable, such as those in early-successional and other transient habitats. Therefore, opportunistic behavior, while a common strategy in pioneer plants, is not necessarily restricted to them. All plants that commonly live in habitats where resource availability is highly seasonal are likely to behave opportunistically with respect to that specific resource. For example, some herbs in the understory of temperate deciduous forests in the eastern United States quickly take up nutrients, especially nitrogen, as soon as they become available during the spring thaw. By this opportunistic behavior, which occurs on a short time scale, spring herbs play an important role in preventing loss of nutrients

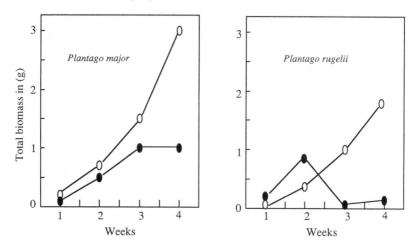

Fig. 10.2. Response of *Plantago major* from repeatedly disturbed habitats (left) and *Plantago rugelii* from less frequently disturbed habitats to nutrient pulses supplied either at the vegetative stage (○) or during reproduction (●) (modified from Miao and Bazzaz 1990).

from the ecosystem (Muller and Bormann 1976). Furthermore, many of these species gain much of their annual carbon when light is plentiful, before canopy closure.

Opportunistic uptake and usage of nutrients has been documented in some early-successional plants, but there has not been a thorough comparative investigation of this strategy. It is not known, for example, how different species in the community differ in their degree of opportunism. Using two colonizing, short-lived perennial congeners common in habitats with different disturbance regimes, we tested the prediction that the species from the frequently disturbed sites (*Plantago major*) would respond to nutrient pulses more sensitively (in terms of growth and reproductive allocation) than *Plantago rugelii*, a species from relatively more stable, late-successional sites (Miao and Bazzaz 1990). Indeed, *Plantago major* flexibly increased its relative growth rate and reproductive allocation with nutrient pulses. Even when nutrient pulses were given during its supposedly deterministic reproductive stage, the plant produced new leaves, greatly increased fruit biomass, and slightly increased mean seed mass and seed nitrogen content. In contrast, *P. rugelii* exhibited limited adjustment in growth and biomass allocation in response to nutrient pulses, and there were no detectable differences in response when plants were pulsed during either the vegetative or reproductive stage (Fig. 10.2).

Nutrient additions in some natural ecosystem do not always demonstrate the plant's opportunistic behavior, even in fast-growing early-successional plants, especially tropical trees such as *Cecropia* (e.g. Harcombe 1977). Competition for nitrogen with soil microorganisms can be intense, and their competence in uptake and use relative to higher plants may greatly reduce nitrogen availablility, despite increased nitrogen supplies to the soil system (e.g. Vitousek 1982). Thus, it is the availability of a resource to the plants rather than its quantity in the soil system that is relevant to plant response.

Another manifestation of the response of early-successional plants to variability in resource availability in their habitats is the speed of their recovery from soil moisture limitation. In pioneer plants, even when their water potential is quite low, recovery from water stress has been observed to occur soon after rain or the addition of water to experimental plots. This fast recovery is generally associated with a low resistance to water flow in early-successional plants. In controlled environments, severely water-stressed individuals of the early-successional annual grass *Setaria* (-3.0 MPa) begin to recover within minutes after the addition of water. Their photosynthetic rate recovers fully within 24 hours (Wieland and Bazzaz 1975). In the field, the placement of the root system within the soil profile can cause differences in recovery among neighbors. After rain, shallow-rooted species respond more quickly than more deeply rooted ones, as the water percolates downward through the upper portions of the soil. In *Setaria faberi* that has many shallow roots, leaf water potential changes substantially from its pattern a day before rain and a day after rain (Fig. 10.3(*a*),(*b*)). In contrast, the deeper-rooted *Abutilon* found in the same patch does not show much change in leaf water potential within the same period. However, if copious water is supplied during rainstorms, water potential in both species recovers at about the same speed. Apparently, both species are equally opportunistic when water reaches their roots.

Plasticity and acclimation

Response *flexibility* refers to the ability of a genotype to function in a variety of environments (Thoday 1953). It results from the plasticity of some traits and the stability of others (Bradshaw 1974, Caswell 1983). Since it is implicitly assumed that flexibility is adaptive in variable environments, the degree of flexibility is best assessed by the growth advantage conferred on the individual having plastic and acclimatory capabilities relative to its performance in the absence of plasticity and acclimation. Two components

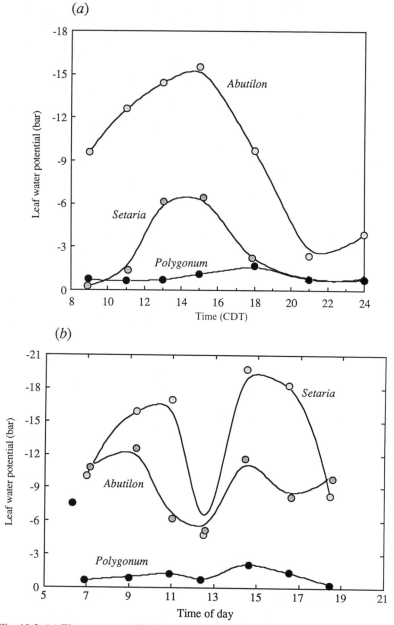

Fig. 10.3. (*a*) The response of leaf water potential of the shallow rooted *Setaria*, the deeper rooted *Polygonum*, and the middle rooted *Abutilon* to a rainstorm after a dry spell in the field. (*b*) The response before the rain storm.

of response flexibility in plants have been recognized: phenotypic plasticity and acclimation. *Phenotypic plasticity* refers to the ability of traits in a given genotype to express different phenotypes in different environments (Bradshaw 1965). Plasticity is adaptive when that expression confers an advantage to the genotype that expresses it in the particular environment (Schlichting 1986, Sultan 1987). *Acclimation* is the ability of a genotype to change the biochemical, physiological, and morphological characteristics of its already established modules in response to a change in the environment. However, because plants grow by the addition of new modules, and since these modules may develop in different environments, plasticity and acclimation are best viewed at the module rather than the whole-plant level. Thus, a genotype that finds itself in a given environment, such as deep shade or an open habitat, and develops a phenotype suitable to that particular environment, is expressing plasticity. In the shade it develops a phenotype which has the anatomical physiological characters that adapt it to shade: e.g. large, thin leaves with lots of chlorophyll. Whereas in the sun, it would have smaller, thicker leaves with much more *Rubisco*. In contrast, an individual that changes the physiology of its modules in response to a change in the environment in which it developed is acclimating to the new environment. For example, leaves on many species are born in the sun but, especially in fast-growing individuals, spend a good deal of their lives in the shade. Since sun leaves and shade leaves do differ greatly in morphology and physiology, acclimation from sun to shade is common in these plants. The ability to adjust quickly to the change, in this case, in the light environment, would be highly adaptive. Acclimation can only involve already established modules. Since already established modules can differ in their ability to change depending on their age (younger modules are more capable of change than older ones), and since the environment may change through the life of many plants, the distinction between plasticity and acclimation is not always clear. The full expression of flexibility in plants may be hindered by a number of carry-over effects at the physiological level or by development traits (Ferrar and Osmond 1986, Woodrow and Mott 1988, Sims and Pearcy 1991). Studies of the flexibility of plant responses to changing environments more often consider plasticity rather than acclimation. As a result, investigators have generally overestimated the ability of early-successional plants to adjust their response, especially their assimilation rates. Acclimation studies more clearly demonstrate carry-over effects that could reduce the magnitude of response to the changing environment.

Because plasticity is a general character of all plants (Bradshaw and Hardwick 1989), the question is not whether pioneer and late successional plants are plastic or not, but how they differ in the degree of plasticity of their traits. It has been hypothesized that, because they occupy highly variable environments, early-successional plants are more flexible than late-successional plants, which are assumed to occupy relatively less variable environments (Bazzaz 1979). The bulk of experimental evidence accumulated over the last decade on both temperate and tropical plants supports this hypothesis.

Much research has focused on plasticity and acclimation to the light environment, especially with respect to photosynthesis and associated physiological and morphological attributes of early-successional plants. Switching plants between contrasting environments, such as high and low light, has been the most common technique of studying acclimation in successional plants. (e.g. Bazzaz and Carlson 1982, Oberbauer and Strain 1985, Fetcher *et al.* 1987, Pearcy 1987, Walters and Field 1987, Popma and Bongers 1988, 1991, Strauss-Debenedetti and Bazzaz 1991, 1996, Newell *et al.* 1993). Seedlings and saplings in deep shade, in intact forests, can suddenly find themselves exposed to full light when a gap is created above them by various agents. Carbon gain and allocation are the most commonly studied aspects of plastic response of plants to changing environments. Rapid changes in structure (e.g. leaf thickness) and biochemistry (e.g. *Rubisco* concentration and activity, chlorophyll concentration, and light saturation) have been observed in plants transferred from low to high light. In some species, however, already established leaves are dropped and a new complement of leaves is formed upon transfer. The emphasis in most acclimation studies was on parameters associated with light capture and carbon fixation. However, flexibility in response to a changing environment involves more than the plant characteristics associated with photosynthetic response to light. Flexibility in allocation to various organs and activities is critical to plant performance in general. It is now well established that various traits of a species may exhibit different degrees of plasticity (Lechowicz and Blais 1988, Bazzaz and Wayne 1994). For example, in the early-successional annual *Abutilon*, the degree of plasticity of different traits in response to light was highest for plant height and decreased in the following order: specific leaf area, allocation to leaf area, number of nodes, allocation to roots, allocation to stems, and allocation to leaf weight (Rice and Bazzaz 1989*b*). In both early- and late-successional species of birch, *Betula*, branch number was the most plastic and the chlorophyll a/b ratio was the least plastic among 18 characters studied (Wayne and Bazzaz 1993*a*).

As mentioned earlier, plants of different successional status differ in their degree of plasticity and acclimation. In a survey of 15 species from early-, mid-, and late-successional plants, Bazzaz and Carlson (1982) found that the differences between the light saturation curves of individuals grown under canopy shade and those growing in the open were much higher in early-successional than in late-successional plants, with mid-successional plants intermediate (Fig. 10.4). Koike (1988) made an extensive study in temperate forest succession in Japan and reached similar conclusions. The bulk of the experimental evidence supports this hypothesis. However, there are some apparent exceptions. For example, Osunkoya and Ash (1991) found constraints on potential acclimation to a changing light regime in seedlings of six late-successional rainforest trees in Australia. They also found that acclimation to increasing light availability was faster than acclimation to decreasing light conditions (Denslow *et al.* 1990). Available evidence suggests that acclimation in early-successional plants occurs by both physiological and morphological change, whereas in late-successional and primary forest species, there is some physiological acclimation but little morphological acclimation ability (Kwesiga *et al.* 1986, Strauss-Debenedetti and Bazzaz 1991, Thompson *et al.* 1992*a,b*).

However, in many successions, the interpretation of flexibility, plasticity, and acclimation are confounded by a shift in life forms during succession: from herbs to shrubs to trees. These shifts in growth habit can make some of the comparisons of the degree of flexibility along successional gradients inappropriate. To avoid this complication, we compared only herbaceous species from an early-successional field, a mature-phase grassland and a mature deciduous forest located adjacent to each other. The species were grown in high and low light conditions and the plasticity of several characters in the two environments was calculated. Plasticity advantage values, estimated from photosynthetic light response curves, in decreasing order, were as follows: early-successional field > prairie > late-successional forest. When plasticity advantage was calculated using both relative growth rate and photosynthetic light response, the same ranking was obtained (Rice and Bazzaz 1989*b*). This test shows that predictions about early- and late-successional plants with regard to flexibility are robust and the patterns are well established. The degree of response flexibility may not be constant, but changes with the age of the individual. In the mid-successional herbs *Aster* and especially *Solidago*, phenotypic variation for size and architectural characters among several populations and families was rather small (Schmid and Bazzaz 1990). However, individuals derived from rhizomes, grown simultaneously in the same environment as seedlings, had relatively higher phenotypic plasticity than individuals derived from seed.

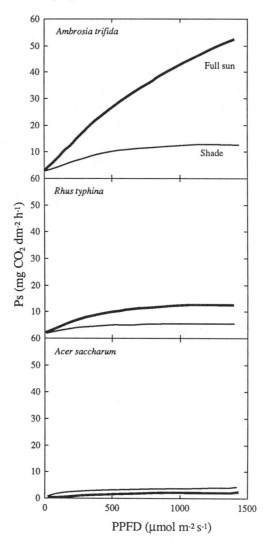

Fig. 10.4. Photosynthetic response to light intensity for early-, mid- and late-successional plants grown in full sun and in deep shade (modified from Bazzaz and Carlson 1982).

This suggests that in clonal plants there may be a change in plasticity associated with the age of the clone. It is not clear how this behavior is adaptive, except if the clone lives for a long time and the environment in which it lives is rapidly changing. Recently, interest in the creation of tree-fall canopy gaps, which suddenly expose seedlings to much higher

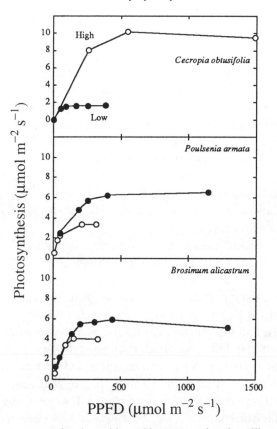

$$\text{PPFD } (\mu\text{mol m}^{-2}\text{ s}^{-1})$$

Fig. 10.5. The response of early-, mid-, and late-successional seedlings of the family Moraceae to conditions during growth in high light and low light (modified from Strauss-Debenedetti and Bazzaz 1991).

levels of light than those under which they developed, has been a major impetus for studies in acclimation and plasticity. For example, there has been much work on the acclimation potential of early- and late-successional tropical plants (review in Strauss-Debenedetti and Bazzaz 1996). Most of these studies compared species from different genera and families.

In order to test these broad-scale conclusions with regard to succession, we investigated whether they also hold true for closely related early- and late-successional species. We used five species in the *Moraceae* family, which differ in their successional status but are a part of the same sere as well as the same phylogenetic unit (Fig. 10.5). In agreement with the general predictions about early- and late-successional plants, *Cecropia* and *Ficus* had the highest photosynthesis and conductance, while the late-successional

Brosimum and *Poulsenia*, and especially the undercanopy *Pseudolmedia*, had lower rates (Strauss-Debenedetti and Bazzaz 1991). We found no acclimation on the basis of biomass allocation, but some degree of acclimation in relative growth rate. In contrast, in a field experiment with rooted cuttings of three *Miconia* species differing in growth, form, and habitat preference in the Costa Rican rain forest, Newell *et al.* (1993) found no difference among the species in the way they respond to gap creation; photosynthetic capacity increased in all species. Furthermore, within the genus *Shorea* that is a major genus in the rainforests of Southeast Asia, A. Moad and F. A. Bazzaz found that the species differ in their growth response to increased light levels. Relative growth rate in the early-successional gap species *Shorea leprosula* responded more positively than did the mature phase *Shorea maxwelliana*.

Plasticity can also be assessed with regard to several traits and resource gradients for various genotypes of a population. Experiments on a wide range of light, moisture, and nutrient gradients with the annual *Polygonum persicaria*, a common species in repeatedly disturbed early-successional habitats, showed that each genotype expressed a set of physiologically, allocationally, and morphologically diverse phenotypes in different environments (Sultan and Bazzaz 1993*a,b,c*). In general, however, the genotypes adjusted appropriately to poor light, moisture, and nutrient conditions by allocating biomass preferentially to those organs that acquire the most strongly limiting resource. For example, in low light conditions, all genotypes allocated more to leaf biomass, and in poor moisture and nutrient conditions, they allocated strongly to root biomass. Because phenotypic plasticity enables a genotype to express diverse phenotypes appropriate to the environment that elicits them, it can confer tolerance of a wide range of environments, and thus increase ecological breadth. This broad phenotypic plasticity may thwart selection and help maintain these genotypes in the population. In terms of the evolution of life history, organisms that exhibit developmental, physiological, morphological, and /or behavioral plasticity and a broad tolerance range are typically considered 'generalists,' compared with ecologically narrow 'specialists' (Baker 1965). These designations draw a conceptual linkage between flexibility in response and niche volume (Chapter 8).

Despite much progress in understanding response flexibility and its patterns in plants, several critical issues related to flexibility remain largely unexplored. Most studies of acclimation compare traits of plants of the same age, rather than those of the same size. However, high levels of resources not only stimulate phenotypic adjustment to the environment, but also promote growth in general. It may therefore be more accurate to compare traits of plants at the same size rather than at the same age (Rice and Bazzaz 1989*a*). For example, when plants of two phenotypes (sun and shade) were reciprocally transferred, plant heights compared at the same age were very similar, but were different when compared at the same plant weight. Thus, light intensity influences plant height independently of its influence on plant weight. In contrast, light intensity influences leaf weight allocation mostly by influencing plant weight. Transferring plants of the early-successional species from low light to high light environments greatly stimulated vegetative growth. In fact, because of the higher initial leaf area ratio, low light plants grew more when transferred to high light, relative to plants which were kept in high light. A similar response was observed in the tropical early-successional trees *Cecropia* and *Ficus* (Strauss-Debenedetti and Bazzaz 1991).

In addition, studies of acclimation and plasticity in plants have considered the biomass advantage conferred by these behaviors. However, there is little information about the fecundity value of these responses. While it is generally assumed that the relationship between growth and overall fecundity is positive, there are circumstances where the two may be decoupled. This decoupling of vegetative growth from reproduction can have significant population-level consequences. In *Abutilon*, plants transferred to high light produced fewer seeds than those that were always in high light (Rice and Bazzaz 1989*b*). Thus, the phenotypic response to increased light can be considered adaptive phenotypic plasticity in terms of vegetative growth, but the decline in seed production suggests that this phenotypic plasticity may not confirm benefits for fecundity.

Finally, it is not known what the cost of flexibility in response is or whether there is any cost to flexibility. Evidence from early-successional plants suggests that little energetic or resource costs exist for plasticity in these plants (Sultan and Bazzaz 1993*c* and references therein). In fact, there is every reason to think that there is selection for phenotypic plasticity in plants (Bradshaw and Hardwick 1989, Sultan 1993), especially for those that live in highly variable environments, such as early-successional plants. It is possible that in these situations the energetic cost of being plastic can be overcome by the large fitness advantage which plasticity confers. These three areas – appropriate comparisons, fecundity advantage and costs – need further investigation in plants in general.

11
Physiological trends of successional plants

The environment of early-successional habitats is open, sunny, and, for many resources, measurably more variable than that of late-successional habitats (Chapter 3). These characteristics of the environment (especially patterns of resource availability), since they have been encountered by the plants through their evolution, dictate, to a great measure, their life history designs and the general physiological behavior. Over the last few decades a wide range of sensitive and accurate instruments have been developed to precisely characterize many of the important parameters in the plant's environment in the field (Pearcy *et al.* 1989). However, as mentioned earlier, the dilemma remains unresolved that this 'measured variability' in the environment, in many cases, does not seem to directly translate into variability in plant response. In a phytocentric view, the plant itself, and not the measured quantity of the environmental factor *per se*, determines the extent of variability. Thus, for some species, a given level of variability may be of no consequence if that variablility is within the range of the breadth of response of these species, but for others, the same level of environmental variability can be outside their range of tolerance and therefore may be detrimental to their growth and fitness. Clearly species from distinct habitats differ in their response to the environment, but, even within the same community and a given patch, similar levels of variation of an environmental factor may elicit different responses from co-occurring species (Fig. 11.1). Do we then assess environmental variability by continuous measurements with precise instruments, or do we use plants of known genotypic identity as 'phytometers,' which by their performance can assess and integrate environmental variability with regard to specific characters or to differential fitness?

Although plants within the same community can differ appreciably in the way they respond to their environment, they do share some common

206

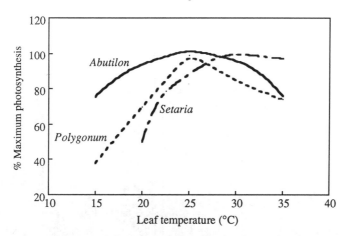

Fig. 11.1. The photosynthetic response of *Setaria*, *Abutilon*, and *Polygonum* to leaf temperature (modified from Wieland and Bazzaz 1975).

features from which generalizations can be made. In fact, unless the habitat is very strongly patchy and the genotypes of a population are highly specialized on these patches, some degree of similarity in response is a prerequisite for co-occurrence in plant communities. Species greatly dissimilar in response to the physical environment are very unlikely to be found in the same location. Based on the available literature and logical deduction, several years ago we made predictions about the physiological attributes of plants of different successional positions (Bazzaz 1979, Bazzaz and Pickett 1980). We suggested that successional plants share many physiological features, including aspects of seed germination, gas exchange properties, growth, allocation, and response flexibility. Furthermore, many of these attributes are shared among various groups, whether we compare early-successional annuals or trees with late-successional plants in both temperate and tropical successions. Extensive studies during the last several years have corroborated most of these predictions in both temperate and tropical regions (see Huston and Smith 1987, Bazzaz 1991*b* for recent reviews).

It should be recognized that some of these predictions consider characters known to be correlated with each other. Therefore, studies have shown that knowing the trends in a given character, we can predict trends in other correlated characters without necessarily having experimental data on that specific character. Additionally, certain sets of characters may be specifically and strongly related to certain environmental factors. For example, many photosynthetic-related characters are highly influenced by the light levels

Physiological trends

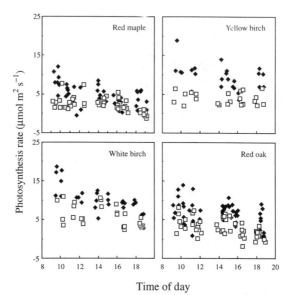

Fig. 11.2. *In situ* photosynthetic rates of upper canopy and lower canopy leaves of red oak, red maple, yellow birch, and white birch during 12 June 1992 in a deciduous forest in central Massachusetts, USA. ■, upper canopy; □, lower canopy. (Bassow and Bazza unpublished)

under which the plants usually grow. Thus, early-successional plants share many of the behaviors of sun-adapted species, and late-successional plants share many of the behaviors of shade-tolerant species, at least during some phases in their life. However, many late-successional and mature phase plants are exposed to conditions of bright light in the canopy, but as seedlings, saplings, and juveniles may spend much time in the shade. In these species, photosynthetic rates and other growth-related attributes of shaded seedlings differ from those of their mature parents. Furthermore, as mature individuals, the microenvironment experienced by the part of the plant at the canopy level can greatly differ from those experienced by the shaded parts of the individual. For example, in the mid- to late-successional *Quercus rubra* (red oak) in temperate deciduous forests, the photosynthetic rates of leaves at the top of the canopy may be as high as those of early-successional plants, while leaves in the shade have a much lower photosynthetic rate, similar to those of late-successional plants. Moreover, morphological and physiological differences between seedlings and adults, and between exposed and shaded parts of individuals, can be greater in some species than in others (Fig. 11.2).

Early- and late-successional plants also differ in responsiveness to increasing light availability. It has been shown that early-successional plants usually increase their photosynthesis greatly when transferred from low light to high light. When the canopy opens, creating a more early-successional habitat, early-successional plants strongly respond. Late-successional plants, on the other hand, show only a limited increase in photosynthesis in response to increased light levels. In this regard, mid-successional plants tend to have intermediate attributes between those of early- and late-successional plants.

But despite the presence of many shade adaptations in the physiology of late-successional plants, e.g. thin leaves, high chlorophyll content, low Rubisco activity, and low light saturation point (Bazzaz and Sipe 1987), it is now recognized that individuals of almost all species in the understory or undercanopy can ultimately benefit from increased light, and that small gaps are necessary for these species to reach the canopy. Furthermore, sunflecks can play a significant role in the carbon economy of many late-successional understory individuals that are found in deep shade (Pearcy *et al.* 1994). In the temperate forest, it is thought that the late-successional shade-adapted sugar maple, *Acer saccharum*, requires a few gaps before it ascends to the forest canopy (Canham 1985). Although late-successional plants can persist in the understory, they do differ in the extent of their persistence. Some species can remain suppressed for a very long time. These species usually have low photosynthetic rates even when they reach the canopy, where many of their leaves are exposed to full sun. Apparently, constraints for long persistence in the understory early in life may preclude full adaptation to high light levels at maturity. Research has shown that there are trade-offs between persistence in the shade and adaptation to take advantage of a high light environment. Therefore, the species does well in one environment but usually not in both. There are, however, a few species, such as *Liriodendron tulipifera* in temperate forests, *Dryobalonops* in rain forests of Southeast Asia, and *Ceiba pentanda* in the neotropics, that can do very well in both environments. It is therefore likely that these species are recruited early in succession when the light levels are high and persist into the mature-phase, late-successional forest.

Extensive studies on seed germination, photosynthesis, growth, and allocation have been carried out by several investigators on a wide range of successions. Despite the differences among various systems, some generalizations can be drawn about succession. Let us look at the physiological characters of early-, mid-, and late-successional plants. We should bear in mind that two common and critical characteristics of the physiology of early-successional plants are: (a) they are relatively short-lived, and (b) they initially occupy open, sunny habitats. Because they are generally transient

Table 11.1. *Some general physiological attributes of early- and late-successional plants*

	Early-successional	Late-successional
Light saturation intensity	High	Low
Light compensation point	High	Low
Efficiency of carbon gain at low light	Low	High
Photosynthetic rates	High	Low
Respiration rates	High	Low
Transpiration rate	High	Low
Stomatal and mesophyll conductances	High	Low
Resistance to water transport	Low	High
Allocation flexibility	High	Low
Acclimation potential	High	Low
Resource acquisition rates	High	Low

in any location, their life history strategies are designed so that they germinate, grow, reproduce, and disperse relatively quickly (Table 11.1).

Seed germination

Seed germination is perhaps the most extensively studied aspect of plant life history. There are hundreds of papers on this subject (see Fenner 1985). Therefore, patterns can be easily detected and generalization confidently made. Extensive evidence from temperate and tropical systems (Bazzaz 1979, Fenner 1985, Vázquez-Yánes and Orozco-Segovia 1993) strongly indicates that seed germination in most early-successional plants is enhanced by light and by fluctuating temperatures, and is much reduced in environments with a low red/far-red ratio, such as intact vegetation. Therefore, seed germination of many early-successional plants is enhanced by disturbances, that destroys the canopy above, changes the quantity and quality of light, and disturbs the soil and brings some seeds with enforced dormancy from deep in the soil closer to the soil surface where temperature fluctuates and the light environment is suitable (Bazzaz 1983). It should be emphasized that, while there may be a few exception to these generalizations, they are based on extensive data from a wide range of habitats and are unlikely to change much with further research.

Photosynthesis

The rapid growth and reproduction of early-successional plants is usually associated with copious leaf turnover and high photosynthetic rates (Reich

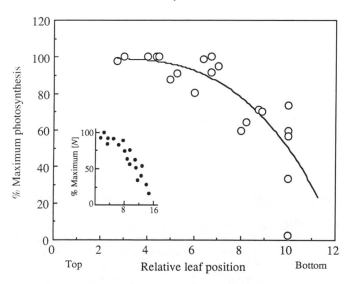

Fig. 11.3. Decline of photosynthetic rates of early-successional annuals with age. Nitrogen concentration also declines with age (modified from Mooney *et al.* 1981).

et al. 1992, Ackerly and Bazzaz 1995*a*). In these fast-growing plants, especially those in nutrient rich mesic environments, leaf turnover rates increase and photosynthetic rates of individual leaves often decline rapidly, as competition for light because of self-shading intensifies with plant age. Nitrogen is quickly reinvested in developing leaves at the top of the canopy, which, being located in a high light environment, are more efficient carbon gainers (Fig. 11.3; Mooney *et al.* 1981, Field 1983, Hirose and Werger 1987). Self-shading plays a major role in leaf senescence and in canopy structure dynamics. While there may be variation in the longevity of leaves on early-successional fast-growing species, there is a negative correlation between fast growth and leaf longevity (Reich *et al.* 1992). In a high light environment the tropical pioneer tree, *Heliocarpus appendiculatus*, had low leaf longevity and a faster leaf turnover rate compared to leaves on shaded plants. Leaf longevity was negatively correlated with photosynthetic capacity (Ackerly and Bazzaz 1995*a*). In summary, it is predicted (Bazzaz 1979) that early-successional plants have high stomatal conductance, high leaf Rubisco content, low resistance to water transport through the vascular system, and high transpiration rates relative to mid- and late-successional plants. These attributes are associated with or are necessary to the achievement of high maximum photosynthetic rates. Extensive research during the last 15 years strongly supports these

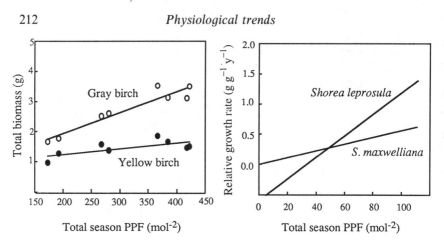

Fig. 11.4. Total biomass in relation to light in the early successional gray birch (*Betula populifolia*) and the later successional yellow birch (*Betula alleghaniensis*) (from Bazzaz and Wayne 1994), and relative growth rates versus light in the early-sucessional *Shorea leprosula*, and the late-successional *Shorea maxwelliana* (Moad and Bazzaz unpublished data).

predictions (see Strauss-Debenedetti and Bazzaz 1996). It has also been shown that these generalizations are robust and can be made, even across families and closely related species, in genera in both tropical and temperate environments. For example, the early-successional gray birch displays a higher maximal photosynthetic rate than the late-successional yellow birch. Other North American *Betula* species, such as white birch and black birch, may be placed on this successional continuum (Wayne and Bazzaz 1993*b*; Fig. 11.4). Similarly, species of the genus *Shorea* in Southeast Asia differ predictably in photosynthetic characteristics; the 'heavy hardwood' species, which are late-successional, tend to show lower photosynthetic capacities, slower growth, and denser wood than do the 'light hardwood' species, which are early successional (A. Moad and F. A. Bazzaz unpublished). Analogous patterns have also been documented within species of the family Moraceae in the new world tropics (Strauss-Debenedetti and Bazzaz 1991, 1996). Such comparative studies of physiological characteristics within plant genera or families provide more stringent tests of adaptive trends by controlling for evolutionary divergence.

Overall, the wealth of available data from our work and various other sources show that the rate of photosynthesis per unit of leaf area generally declines with succession and that photosynthetic rates of early-successional annuals can be as high as $50\,\mu\mathrm{mol\,m^{-2}\,s^{-1}}$. Plants in the understory of mature phase forests usually have very low photosynthetic rates, often as low as 1–$2\,\mu\mathrm{mol\,m^{-2}\,s^{-1}}$. The commonly observed steepness of the initial

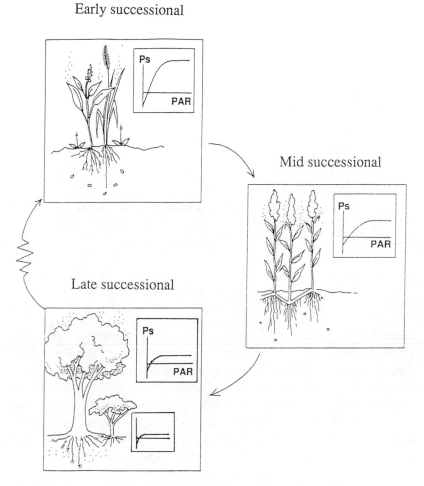

Early successional

Mid successional

Late successional

Fig. 11.5. Carbon dioxide exchange attributes of early-, mid-, and late-successional plants in relation to light.

slope of photosynthesis with increased light means that late-successional trees are photosynthetically more efficient at low light intensities than are early-successional herbs. Mid-successional trees generally have values between early-successional herbs and late-successional trees (Fig. 11.5). However, such patterns in photosynthetic light responses are dependent on the units in which photosynthesis is expressed, and may commonly be most pronounced on a leaf dry weight or leaf nitrogen basis (Givnish 1988).

Despite clear differences among species within each group (Pacala *et al.* 1993), canopies of early-successional trees, especially in the tropics, are

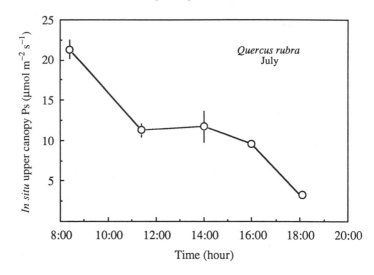

Fig. 11.6. Decline in photosynthetic rates of canopy leaves of red oak over the course of the day (Bassow and Bazzaz unpublished).

generally more open than those of late-successional trees (Horn 1971). This allows more light to pass through their canopies. But regardless of the degree of openness and availability of high light, maximum photosynthetic capacity (P_{max}) is often not attained, even in early-successional plants. Limitation of resources such as water and nutrients, herbivore damage, and internal negative feedback mechanisms, such as sucrose and starch accumulation in chloroplasts after an active period of photosynthesis, can cause a decline in photosynthetic rate. For example, leaves of the upper tree canopy often reach maximum daily photosynthetic rates in the early hours of the day and gradually decline over the course of the day (Fig. 11.6). Furthermore, a midday drop in photosynthesis may occur, caused by stomatal closure in response to a high vapor pressure deficit or the limitation of soil water supply (Schulze 1986). Such a midday depression, although varying in magnitude in each case, can be observed in both early-successional species and the exposed canopies of late-successional plants. Accessibility of soil water, plant resistance to water transports, and stomatal sensitivity to vapor pressure deficit and internal CO_2 concentration (C_i) dictate the pattern of daytime patterns in gas exchange rates. But knowledge of diurnal patterns in photosynthesis and transpiration of emergent and canopy trees in mature-phase forest is extremely limited. Portable gas analysis systems and canopy access techniques (e.g. towers, cranes, cherry pickers, and canopy platforms attached to balloons) will greatly help in this regard.

Despite the presence of mechanisms that adjust photosynthesis in response to the degree of leaf exposure to varying light levels, the rate of photosynthesis usually remains unchanged if the leaves of early-successional plants are exposed to light intensities above their saturation point. That is, they are not expected to exhibit 'photoinhibition.' It is likely that these plants have developed biochemical mechanisms to dissipate excess energy, and therefore suffer no damage in high light. Many seedlings and saplings of late-successional individuals, in contrast, may show much reduced photosynthetic activity or even mortality when exposed to high light intensities due to excessive heat loads on the leaves and photoinhibition. A third response to sudden increases in light intensity is seen in some herbs in the understory of mature-phase forests, such as *Asarum* (Bazzaz and Carlson 1982), that quickly lose their leaves when exposed to high light and produce a new complement of sun-adapted leaves with the appropriate morphology and physiology.

Under certain circumstances, other characteristics not directly related to successional position may also be of considerable importance as predictors of some of these physiological attributes. For example, some mature-phase multistory rain forests in Southeast Asia usually have only a small number of early-successional tree species. It has been found in these systems that species' stature may account for most of the variation in photosynthetic physiology (Thomas and Bazzaz unpublished data). Leaves of species with higher asymptotic maximal heights have higher photosynthetic capacities (Fig. 11.7).

Transpiration

Because stomatal conductance controls CO_2 and water vapor exchange between the leaf and the surrounding air, a positive relationship between photosynthesis and transpiration is expected. Because of high stomatal conductance in early-successional plants, transpiration rates are generally high as well. In contrast, late-successional plants generally have low rates of transpiration, about one-quarter of that of early-successional plants on a leaf-area basis. These differences in rates of transpiration and photosynthesis reflect differences in both stomatal conductance and Rubisco concentration and activity, both of which generally decrease with succession. (Like photosynthetic rates, transpiration rates expressed on a leaf area basis alone do not give complete information on the plant's carbon gain or water

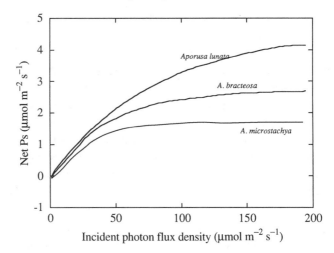

Fig. 11.7. Photosynthetic light response curves in congeneric *Aporusa* differing in maximal attainable heights in tropical rainforests (from Thomas and Bazzaz unpublished data).

loss patterns. They must be scaled to the entire leaf area of an individual, taking into account differences among leaves in these activities in relation to age, position in the canopy, the level of herbivory, etc.)

Energy balance considerations (Gates 1965, Nobel 1991) of both open and shaded environments suggest that high stomatal conductance and the resulting high transpiration rates in early-successional species may be effective in preventing superoptimal leaf temperatures when there are high radiation loads, as much energy can be dissipated as latent heat of evaporation (LE). For example, under field conditions, we found that the early-successional *Abutilon* can maintain leaf temperatures near 28 °C (near optimum for photosynthesis) even when air temperature climbs to 35 °C. It is also possible that the low leaf conductance of shaded, late-successional seedlings and understory species may prevent unnecessary cooling. Currently, however, there are too few comparative studies to discover if early- and late-successional plants or their shaded or exposed parts optimally 'solve' the energy budget equation. It is not known if early-successional plants in open environments behave similarly to the exposed canopies of late-successional plants in this regard. We do know, however, that the pioneer trees in gap in the tropical rain forest of Mexico display their leaves non-randomly such that they receive most of their light from diffused rather than from diverted beam radiation (Ackerly and Bazzaz 1995*b*). We do not know if plants in early- and late-successional environments dissipate heat differently or how they change their Bowen ratios (the loss of heat by sensible heat transfer (H) relative to the loss of heat by LE), according to field conditions. We also do not know if the cooler soil water passing rapidly through plants with high transpiration rates plays a role in energy dissipation from leaves, especially on warm days in the spring and early summer. Clearly, there is much research to be done in this area.

Respiration

Largely because of technical difficulties, whole-plant respiration has not been extensively studied in successional plants, and leaf respiration *per se* is of limited explanatory value in understanding the carbon economy of plants. Root and stem respiration are especially critical to assessing net carbon gain in plants. While plants gain carbon mainly by leaves during the daytime hours, the plants respire both during the day and the night from all their living parts. They can also lose much carbon in root exudates, both to herbivores and to support symbionts such as nitrogen fixers, bacteria in nodules, and mycorrhizae. On a leaf area basis there is generally a decline in leaf respiration with succession. However, the rate of decline with succession of dark respiration on a leaf area basis appears to be less steep than that of photosynthesis (Bazzaz 1979). Furthermore, the stems of many herbaceous early-successional plants are green and photosynthetic. In contrast, stems of late-successional trees are usually not. However, because of the usual increase in respiring over photosynthesizing tissue with size, it is expected that whole-plant respiration will increase with succession. It is, however, unknown what the trends would be if dark respiration rates and photosynthesis were expressed on weight bases. Differences in allocation to belowground parts for early- and late-successional plants can complicate these issues, because of the difficulties in accurately measuring root respiration *in situ* independent from respiration of the rest of the soil.

The rate of dark respiration and the initial slope of photosynthetic response to light determine the light compensation point that indicates the intensity at which photosynthesis balances light respiration. The light compensation point is predictably lower for shade-adapted, late-successional plants than for sun-adapted, early-successional ones (Bazzaz 1979). Operating at the light compensation point results in net carbon loss to the individual because additional carbon is lost during dark respiration at night. Therefore, light compensation points, usually obtained by extrapolation from light response curves of individual leaves or whole shoots, are less informative

Physiological trends

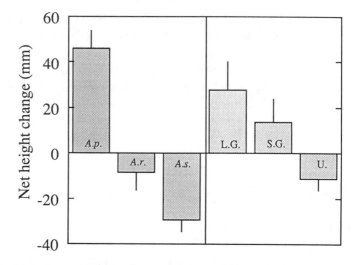

Fig. 11.8. Net change in height of *Acer pensylvanicum* (*A.p.*), *Acer rubrum* (*A.r.*), and *Acer saccharum* (*A.s.*) and of all *Acer* species in large gaps (L.G.), small gaps (S.G.), and in the understory (U.) in a deciduous forest (modified from Sipe and Bazzaz 1995).

than they were originally thought to be. The whole-individual light compensation point, which takes into account daytime whole-plant photosynthesis and daily whole-plant respiration, is more informative about persistence and growth in the understory. In order for plants to survive, grow, and reproduce, they must have a life-long positive carbon gain (Mooney 1972). Many understory individuals of late-successional plants can persist for some time in light-limited environments by hovering near zero carbon gain. Some species can persist for some time even when their net carbon gain is negative. These individuals can lose weight. In the field, we observed that reduction in growth in seedlings and saplings can be brought about by negative carbon balance, and by leader damage by herbivores and severe winter weather (Sipe and Bazzaz 1995; Fig. 11.8). Successful individuals must be able to tolerate these conditions until more resources become available, such as when a canopy opening occurs overhead and year-long net carbon gain is restored. In temperate forests it is not uncommon to find many small late-successional tree seedlings in close proximity in the understory despite the fact that they differ greatly in age; extended periods of zero or negative carbon gain result in individuals of similar size. But trade-offs may exist between long persistence in the understory and fast growth. For example, the late-successional *Shorea maxwelliana* persists in the understory of the rainforest in Southeast Asia despite obvious periods of a net carbon loss. On the other hand, seedlings of the early-successional *Shorea leprosula* cannot survive for long unless they are gaining carbon. But when light becomes available *Shorea leprosula* grows much faster than *S. maxwelliana* (Moad and Bazzaz unpublished).

In order to obtain a deeper mechanistic understanding of carbon gain and allocation in successional plants and how it changes through succession above and beyond general trends, multiple parameters must be measured. There are, however, only a few analyses of whole-plant photosynthesis that consider the total leaf area, leaf life span, age-specific photosynthetic rate, leaf exposure to light, etc., all of which are critical to calculating mechanistically or modeling whole-plant carbon gain, its changes through the season, and its relation to the patterns of the availability of critical resources and its response to herbivory. Similarly, there are even fewer data on whole-plant dark respiration, export and exudation, which close the plant's carbon budget.

Water-use efficiency

Water can be a highly variable environmental factor in successional habitats. Water limitation causes stomata on the leaves to close and reduces carbon gain and cell expansion, and can greatly reduce plant productivity (Boyer 1982). Stomatal sensitivity to changes in leaf water potential differs widely among plants (see Kozlowski *et al.* 1991). Plants also differ with regard to the water potential at which reductions in photosynthesis and transpiration begin, the steepness of their decline, the water potential at which they become negligible, and the water potential from which they can recover. Late-successional plants, especially individuals in the understory, may be generally more sensitive to declining soil moisture levels than are early-successional plants. Available evidence, though limited, is consistent with this prediction. Generally, maximum photosynthetic rates are attained when leaf water potential is least negative, and photosynthesis declines when the leaf water potential drops to a certain level, which can differ greatly among species within the same successional community (Fig. 11.9). However, some species that occupy eroded mid-successional habitats on thin soil, which are exposed to frequent droughts, can actually experience increased photosynthetic rates as the leaf water potential declines from near maximum; photosynthesis then declines as leaf water potential drops further (Bacone *et al.* 1976, Ormsbee *et al.* 1976, McGee *et al.* 1981). It is not yet well established how widespread this behavior is among mid-

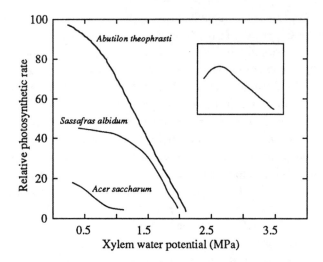

Fig. 11.9. Relationships between photosynthetic rates and leaf water potential of plants of different successional positions. Inset shows the response of mid-successional trees that invade particularly drier sites on thin soil.

successional plants and what the mechanistic bases are that underlie it.

Water-use efficiency (WUE), the amount of water expended in transpiration to obtain a unit of carbon dioxide from the surrounding air, is influenced by the magnitudes of the leaf conductances to water vapor and CO_2 exchange. Water-use efficiency analyses have been mostly instantaneous, the rate of net photosynthesis per unit area per unit time is divided by the rate of transpiration per the same unit area and unit time. Because of changes during a day in stomatal conductance and Rubisco activity and concentration, especially in early-successional species, WUE will depend on the time it is measured and may be highly variable. Season-long water use efficiency undoubtedly differs from instantaneous water use efficiency. In order to more accurately assess the actual WUE and compare species, WUE should be obtained by measuring total biomass accumulation and total water consumed during one time period. This could be done for individual plants or whole ecosystems. We have found in one study that by efficient NO_3-uptake, *Solidago* neighborhoods in mid-successional habitats induce both low water use efficiency and photosynthetic rate of the invading tree seedlings and can greatly delay their entry (Burton and Bazzaz 1995). Unfortunately, there is little analysis of WUE for early- and late-successional plants, even at the whole ecosystem level.

Fast aboveground growth and fast turnover of leaves of early-successional and pioneer plants is expected to be matched by fast belowground growth

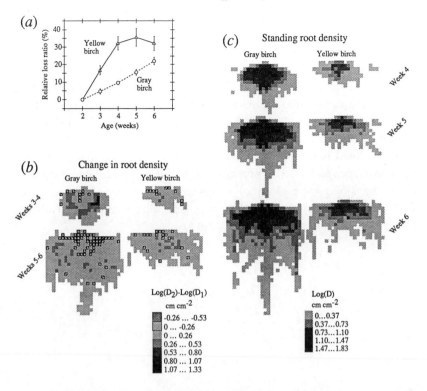

Fig. 11.10. Patterns of root production and loss in gray birch (*Betula populifolia*) and yellow birch (*Betula alleghaniensis*). (*a*) Relative loss ratio through six weeks of development. (*b*) Changes in root density over time. (*c*) Changes in standing root density (modified from Berntson *et al.* 1995).

and turnover of fine roots. However, there is very limited comparative studies on underground activities of plants from early and late successional plants. In a study comparing the early-successional gray birch, *Betula populifolia*, with the late-successional yellow birch, *Betula alleghaniensis*, we found that patterns of root production and loss for the early successional, fast-growing gray birch and the slower-growing, mid-successional yellow birch species demonstrate that the rate, magnitude and intensity of belowground foraging can be highly variable between species of contrasting successional status (Fig. 11.10). *Betula populifolia* produces root systems that fill a larger amount of space more rapidly than *B. alleghaniensis*. However, the relative amount of roots lost to those produced during the development of plant root systems appears to be decoupled from the degree of localization of root loss. Yellow birch showed greater overall root relative loss ratios, while gray birch showed more pronounced localized relative root loss and production (Berntson *et al.* 1995). In gray birch, new roots were produced in spreading concentric bands while older roots near the stem were lost. Yellow birch, on the other hand, distributed its roots within a smaller amount of soil in a more diffuse spatial distribution of relative root loss and production. The net effect of these differences is that yellow birch shows little capacity to expand its range of soil exploration, while gray birch moves its production of new roots rapidly through the available soil.

Allocational flexibility, acclimation potential and resource acquisition rates have been addressed in previous chapters. Our current understanding of the physiological ecology of successional plants suggests the trends described in this chapter. Whether these generalizations apply to a wider range of succession remains to be seen.

12

Crossing the scales: can we predict community composition from individual species response?

Models and predictions

Over the last several decades, plant ecology has been moving from simple description and classification of vegetation to understanding causes and mechanisms from which predictions can be made about community structure, dominance, and diversity, and about the future of different ecosystems. The desire for predictions is being enhanced by the pressing need to appropriately manage ecosystems and to understand how critical aspects of community organization respond to our rapidly changing climate. Predicting the future of populations and communities is currently a major goal in dynamic ecology. Uncertainties about the future of the human environment and the future ability of ecosystems to supply goods and services have greatly encouraged the search for predictive models. Such models are being developed to understand the impact of human activities on biodiversity and other changes in the global environment (Houghton *et al.* 1990, 1992, Lubchenco *et al.* 1991, Wilson 1992). Currently, the scientific community is being challenged to answer questions about the future of major ecosystems under various global change scenarios. These predictions can be of great importance to humankind, as they may be used for the formulation of national and global policy. Predictive models are being used to anticipate community composition under climate change conditions (Pastor and Post 1988, Bolker *et al.* 1995), changes in source-sink patterns of global carbon (Tans *et al.* 1990, 1995, Melillo *et al.* 1993, Smith and Shugart 1993), species migration in response to global warming (Davis 1989, Woodward *et al.* 1991), and many other large-scale phenomena. Forest models are being developed and used to predict how forest structure might be impacted by global change elements such as air pollution,

223

224   Community composition

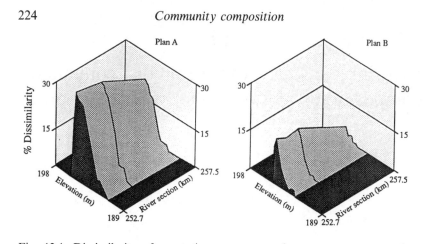

Fig. 12.1. Dissimilarity of vegetation response surface to present vegetation. Predicted dissimilarity would be generated by a modification of the natural flooding regime caused by the implementation of two engineering plans (A and B) for building a dam on a midwestern river. Plan A includes water retention pool at 189.9 masl and plan B includes a pool at 183.3 masl (modified from Franz and Bazzaz 1977).

temperature rise, and elevated CO_2 concentrations. Climate models, such as General Circulation Models (GCM) that predict temperature and precipitation on a regional scale, are being coupled with ecosystem models to produce predictions about future ecosystem productivity, biogeochemical cycling, and other aspects of large-scale ecosystem functions (Melillo *et al.* 1993). Even before the awareness about global change, models played a significant role in physiological, population and community ecology. Various models of ecosystem recovery have sought to describe and predict successional patterns (see Usher 1987). Carbon gain models in physiological and ecosystem ecology (see Reynolds *et al.* 1993), matrix models in population ecology (May 1973, Caswell 1989, Silvertown and Lovett-Doust 1993), and a large number of ecosystem dynamics models in community ecology are now widely used for predictions. Most of these models involve scaling of behaviors and processes. Thus, moving across scales is becoming more common than before. But even simple models can offer predictions that are relevant to decision-making in specific cases.

For example, modification of the hydrologic regime of floodplains can be great when dams are constructed on rivers. A simple probabilistic model was developed based on species habitat preference and used to resolve a controversy between engineering plans and environmental preservation in a floodplain forest dominated by *Acer saccharum, Populus deltoides, Ulmus americana,* and *Celtis occidentalis* (Franz and Bazzaz 1977) (Fig 12.1). The designers of the dam claimed that their proposed designs would have little impact on the vegetation. However, a simple model in which each species was represented by the normal density function based on both field measurements of the three-dimensional coordinates of individuals of the important species and on flooding frequency clearly indicated that all three engineering plans for the dam would have a great impact on the vegetation. The model simulation was designed so that it would give the estimated impact for any design, with the flood stage probabilities taken directly from the engineering plans. The model then produced surfaces of dissimilarity from current conditions. Dam building is currently being practiced very widely, especially in developing countries where demands for water for agriculture and domestic use is rapidly increasing. Therefore, even simple models, such as the one described above, could be of much utility to make predictions about possible impacts on vegetation and may help avert great disasters. The project of building the dam on the river was abandoned because of our study.

Simple scaling in ecosystems

Significant differences exist between population and community ecologists on one hand and ecosystem scientists on the other. For population and community ecologists, individual genotypes and species do matter in the organization, function, and future of these assemblages. In contrast, many ecosystem scientists believe that it may not be necessary to know the details of individual behavior in order to predict system-level response (Foin and Jain 1977, Shugart 1984). The 'big leaf model' that is widely used in ecosystem ecology, for example, does not specifically consider the individual behavior of various species in an ecosystem. It concerns the carbon gain, nitrogen relations, etc. on a whole-system level. At this level, measurements of carbon flux, primary productivity, decomposition, and nutrient release could be made using integrating techniques such as eddy correlations, soil carbon dioxide flux, and watershed flows. These approaches consider the response of a unit of the landscape and integrate the response of all individuals present without regard to differences among them.

Scales in space and time for plants can vary by several orders of magnitude (Osmond *et al.* 1980). Despite universal appreciation of this fact, there remain several complications in understanding how plants simultaneously process energy and resources on widely different scales. Complications arise when communication among scales are not defined and when higher level observations are used to infer lower level processes (Wiens *et al.* 1985). It is instructive to note that both physiological and ecosystem ecology consider similar processes but on a very different scale (see Bazzaz and Sipe 1987). Obvious parallels exist between leaf energy balance and ecosystem energy balance, leaf level carbon gain and system productivity, leaf level nitrogen use and nitrogen biogeochemical cycling, etc. Because of these relations, physiological ecology and ecosystem ecology have been intimately connected. The emerging discipline of ecosystem physiology (Mooney 1991), reflects that intimacy. The emphasis in this field is on understanding dynamics by studying processes, rather than just their outcome. Physiological ecologists are scaling upward, from single leaf photosynthesis and transpiration to whole-plant carbon balance and water use, by considering environmental variability around parts and variability of parts in responding to their changing environment (Schulze 1982, Mooney and Chiariello 1984). Simple plant physiological descriptors, such as stomatal conductance or tissue nitrogen concentrations, are being used as predictors of vegetational distribution and large scale primary productivity (e.g. Schulze *et al.* 1994).

Scaling leaf-level response to canopy-level response is more common in agriculture and forestry. However, it is now more and more practiced in ecology. Plants in growth form can provide the means for addressing the great diversity of physiological responses and allow plant-level physiology to be scaled to higher levels (Chapin 1993). Scaling methods can greatly vary in their level of complexity and the number of parameters and mathematical equations used. However, in many cases simplified equations can be used for scaling to the next higher level (see Norman 1993). For example, parameters as simple as absorbed photosynthetically active radiation (APAR), which are not difficult to measure in the field, and ε_c, the light use efficiency, can be used to predict net primary production (NPP) in many ecosystems using the following equation: $NPP = \varepsilon_c \int APAR \, dt$. Similarly, transpiration from a canopy can be predicted by knowing the remotely-sensed normalized difference vegetation index (NDVI), photosynthetically active radiation (PAR) and transpirational efficiency (ε_w) (Running and Hunt 1993).

Many current ecosystem models make the assumption that all units of photosynthetic ecosystem are equivalent and therefore the producers can be treated as a 'big leaf.' However, in most circumstances in the field there is much heterogeneity in resource availability. Individual plants in an ecosystem can differ greatly in their effect and response to the environment and the whole system is temporally dynamic, i.e. the big leaf is made of patches of differing activities that are changing in time. In addition to environmental heterogeneity, several other factors can compromise the simplicity of models (see Bazzaz 1993). They include incongruent availability of resources and phenotypic variation.

In its beginning, plant community ecology was intimately related to plant geography, and factors of the physical environment that control plant distribution and abundance were emphasized. As the discipline progressed, a more dynamic view emerged and emphasis shifted to understanding the processes that produce the observed patterns of plant communities in nature (Watt 1947). The more recently developed field of evolutionary plant ecology aims at understanding the evolution of specific traits and behaviors that lead to these observed patterns. Functional ecology combines studies of physiological adaptation to the prevailing conditions in the environment with the evolution of these adaptations. The field has thus expanded to include studies of plant systems on all spatial and temporal scales. Unlike studies in animal community ecology that emphasize community organization, many studies in plant community ecology (e.g. Whittaker 1975) have classically emphasized the following:

1. Describing vegetational composition, including species identities, dominance hierarchies and diversity.
2. Identifying community types and their boundaries, either subjectively or by multivariate analysis of data collected from several stands, and relating community types to one another successionally.
3. Relating community attributes to habitat factors by direct observation or by correlation of species composition with environmental gradients.

With the universal recognition that communities are dynamic rather than static, and that various patches within a community can exist simultaneously in different stages of succession, we now seek to know the consequences of disturbance regimes, especially in terms of the modification of resource availability, and to predict spatiotemporal changes in community structure. Therefore, analysis of changing patterns of resource fluxes caused by disturbance, comparative physiological ecology, and the demography of populations are being integrated in the study of plant communities. Modern plant community ecology has come to encompass most research approaches in plant ecology, including environmental measurements, physiological ecology, and interactive demography of the component species. Since these activities and processes can occur at different scale and because their interpretation is scale dependent, much emphasis is being put on how scaling is done (see Ehleringer and Field 1993). Because it is easier to collect and accumulate information about individual species than observe changes in communities over long periods, ecologists often ask

'Can we predict community composition by knowing something about individual species response?' Perhaps more appropriately: 'What must we know about individual species response to the environment to make predictions about the future of the community?'

Unlike physiological and ecosystem ecology, community ecology has confronted middle-number systems, which are extremely difficult to study (see Allen and Starr 1982 for a discussion), and has often relied on observation of state to infer process. Furthermore, community ecology has focused on species relations and their role in community organization, without considering how different genotypes within these species respond to each other or to their environments and what role they play in the future of the community. In contrast, population ecologists emphasize the variation within populations in communities. Many of the forces that maintain species diversity in the community also maintain genotypic diversity within populations in the community. Especially in late-successional, long-persisting communities, genotypic diversity may be as critical for species diversity for the future of that community and its stability in the face of change. It is known that the genotypic structure of populations in communities can also differ in different patches and years, as different genotypes are encouraged by specific changes in the environments. Different genotypes can be found in patches differing in levels of environmental resources due to their different tolerances of these resources. When there is restricted dispersal, sibling genotypes may be found in the same patch, and a succession of genotypes can be recruited in the same patch but in different years, depending on weather patterns and response of various genotypes. In fact, because early-successional plants are usually autogamous, members of full sibling families may be competing with each other in one patch. Until recently, the performance of genotypes in different patches in the field (e.g. Antonovics *et al.* 1987) or on controlled gradients could only be evaluated with perennial plants that could be cloned by separating tillers or rhizome fragments. However, cloning by other techniques such as axillary bud enhancement now allow, even for annual plants, an analysis of the response of individual genotypes to the environment by growing replicate identical genotypes on environmental gradients. Extensive experimental evidence with the early-successional annuals (e.g. Sultan and Bazzaz 1993*a,b,c*, Thomas and Bazzaz 1993) shows that great differences can occur among these families and genotypes of a population in response to the environment. All these spatial distributions of genotypes can have profound effects on how the community as a whole responds to a changing environment. The separation of population biology from community ecology is thus functionally artificial, and predicting the future of a community must also take into account the structure of its populations.

Community-wide predictions

The answer to the question of whether we can make accurate community-wide predictions based on knowledge of certain species response can be deceptively simple. A pessimistic view would assert that communities and their species and environments are extremely complex. Along with chance events, many plant characteristics can play a role in determining community structure, and these characteristics interact with the changing patterns of resource fluxes in the environment. Likewise, initial conditions may play a major role in community trajectories. In such a complex and dynamic situation, it would be exceedingly difficult to predict how these factors interact to generate community patterns, especially dominance hierarchies and diversity trends, and how these patterns change over time and space. An optimistic view, however, would assert that by knowing something about species availability, their dispersal modes, recruitment and growth patterns, competitive relationships, and the general trends in critical environmental factors, we can make predictions about the future of communities. In many cases, exact predictions are perhaps unattainable and may be unnecessary, but general trends are recognizable and useful. Simple models based on a few parameters can inform crucial policy, as discussed above. However, the scale of resolution influences predictions of community structure. The scale of resolution also determines the kind and detail of knowledge required. Does one wish to predict the change in the importance of beech in temperate forests with an incremental increase in temperature and CO_2, or the change in the dominance hierarchy of tree species for one hectare on a north-facing slope in rain forests of Puerto Rico after a hurricane blowdown? Do we wish to predict global carbon storage with a doubling of atmospheric CO_2 levels, or a change in hierarchy of species composition in a grassland affected by a 3 °C temperature rise over a 50 year period? The answers to these questions may dictate how we go about making these predictions. What kinds of data are needed? What approaches do we take? What mathematical models should we use? Currently these questions are not only of scientific curiosity, they are also at the heart of discussions critical to society.

Predicting successional change

Because succession is a dynamic process, predictive models have been an important part of its study. Conceptual, verbal, and box and arrow models

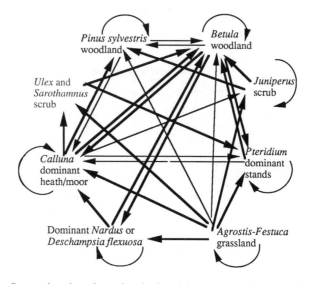

Fig. 12.2. Successional trajectories in heath/moor, grassland, and woodland ecosystems (modified from Miles 1979).

were commonly employed early in the study of succession. In fact, the classical Clementsian view of succession can be seen as a form of a deterministic conceptual model, while that of Gleason can be viewed as a stochastic model (Usher 1987). Most early papers on community description, especially in the United States, included diagrams depicting presumed successional relationships among community types. Almost without exception, classical vegetational studies considered successional relationships and trajectories. Later, transition matrix models were introduced to predict successional outcome (Stephens and Waggoner 1970, Horn 1975, Shugart *et al*. 1981, Usher 1987). By knowing the relationships between community types or species and the probability that one replaces another (transition probabilities), the future of the landscape can be predicted using matrix algebra. Currently available models differ greatly in the level of detail required for making different predictions. Many of these models pertain to forest dynamics, and their application to and linkages with succession in other habitats have not yet been accomplished.

Frederick E. Clements, the most prolific writer on the subject of community succession, as well as many others after him, have used various features of plants and their habitats to predict successional patterns. Many of these 'conceptual' models were driven by a strong belief in the directionality and preordained outcomes of succession. Commonly, different community types within a region were related to each other in successional terms, producing graphical representations of the perceived successional relationships, all culminating in 'climaxes' defined as self-perpetuating communities. These approaches have produced valuable information about community successions (see McIntosh 1980). The simplicity of these models can be deceptive, though, since many of them were based on extensive field observations and entailed a good understanding of many critical life history traits and their role in species replacement (Fig. 12.2) (e.g. the vital attributes of Noble and Slatyer 1980). For example, it has been 'predicted' (and longstanding observations confirm) that, especially in the absence of fire, pine stands in the southeastern United States are followed by hardwoods (oaks, hickories, etc.) because the pines provide the animal dispersers of the hardwood seeds with attractive habitats, the hardwood seeds with suitable germination sites, and the developing seedlings with initially needed shade: this, then, is an accurate prediction based on the knowledge of some life history traits.

But despite their usefulness, many of these predictions about successional change generally use important but still vaguely defined notions such as 'site modification,' 'shade tolerance,' 'competitive superiority,' 'vital attributes,' and recently, 'resource use efficiency.' While these concepts appear simple, they can be very complex and each can encompass a great number of the biochemical, physiological, demographic, and genetic attributes of the species. They need to be better understood in order to achieve a truly mechanistic and precise interpretation of the patterns of successional change in ecosystems. We believe that resource modification, uptake, allocation, deployment and use efficiency encompass critical plant function and that their understanding will be of paramount importance to community organization and dynamics.

Succession models use a wide range of plant attributes in their structure. The most widely known forest succession models, such as FORET-type models, use some physiological and demographic attributes of species to make predictions about the structure of forests (Shugart *et al.* 1981, Shugart 1984). FORET-type models are being extended by the inclusion of additional physiological parameters and used for a wide range of predictions, particularly with regard to issues of pollution and global change (e.g. Davis and Botkin 1985, Leemans 1991, Prentice *et al.* 1993). The resource ratio model (Tilman 1988) uses light and nutrients. It assumes that competitive ability for light and nutrients are inversely related. In this model, interactions occur among populations rather than among individuals and may be especially applicable to succession under limiting soil nutrients. The

Huston–Smith (1987) model, on the other hand, can handle multispecies interactions and multiple resource competition.

It is clear that getting detailed knowledge about a system and its component species can be challenging, time-consuming, expensive, and in some complex systems, currently impractical. What, then, is the alternative? A practical approach to the study of community change and succession would be to reduce the number of environmental attributes considered and species investigated. For example, some environmental variables can be aggregated, using multivariate techniques, to generate eigenvalues and eigenvectors, so that fewer axes account for a significant fraction of the overall response (Bazzaz 1987). Plant species, although they differ somewhat from each other, can also be grouped into 'guilds' (Root 1967) or 'functional groups' (Smith *et al.* 1995), which are essentially aggregations of similarly responding species. Species of a community are grouped together by similar function or similar response to environmental factors. In such cases, information about the group, rather than about each member of the group, may be sufficient to make predictions about their response. These groups are constructed to address specific questions, and their membership can vary depending on the questions to be answered. Which parameters are positively associated with each other? How strong are these associations among variables and among species? Under what circumstances could they break down? Answers to these fundamental questions in community ecology would lead to appreciable progress in the field.

Let us examine the life cycle of a typical plant and identify some obvious attributes that influence individual species response and interspecific interactions. The major attributes in a typical plant life cycle are:

1. *Dispersal*: efficiency, means, and patterns in space and time.
2. *Seed*: dormancy, seed bank longevity, germination, and emergence.
3. *Growth*: relative growth rate (RGR), allocation, allometry, deployment, and survivorship.
4. *Reproduction*: mode, time, frequency, trade-offs, costs, and lifetime fecundity.

What predictions about the structure of a successional community can we make by knowing something about the above stages of the life cycle of its major plant species? What would community structure be in patches in a field differing in moisture, nutrients, light, etc.? What variations in community composition are expected to occur in years of different environmental of conditions, such as severe droughts or occasional extensive defoliation by herbivores? Aspects of these attributes, individually and in various combinations, can allow general predictions to be made. There is little doubt that predictions about short-lived communities such as early-successional ones are easier to make. Knowledge about the life history attributes of their component species can quickly accumulate because of their short life spans and the relative ease of experimentation. (There is, however, no reason to suspect that later-successional, longer lived species cannot yield knowledge about their life history to be similarly used in making predictions.) Based on our extensive knowledge of aspects of the physiological and population ecology of the species of the early-successional annual community gained over several years of field observations and experiments, we can make the following predictions about its structure and how it may change in time and space.

Predicting community composition from life history patterns

Seed germination

It is well known that seed germination can be a major determinant in recruitment time and the development of size hierarchies in plant populations (see Harper 1977, Silvertown and Lovett-Doust 1993). Species or genotypes that are recruited early usually preempt resources in a patch and, through asymmetric competition, dominate the community. We investigated factors critical in influencing community structure in early-successional fields. We found that the response of seed germination of the oldfield annuals is quite similar on a moisture gradient: all species fail to germinate in standing water and in dry soil. Also, we found that the differences in germination among the species in intermediate soil moisture levels are too small to be important in the fluctuating environment of the field. Therefore, it is concluded the response of germination to soil moisture is not very useful for predictions.

In contrast to their similarity on the moisture gradient, the species respond differently to temperature. For example, *Ambrosia* germinates at lower temperatures, while *Setaria* germinates at much higher temperatures (Bazzaz 1984*b*). In the field, the species emerge at different times and from different depths, and their emergence time corresponds well to their temperature requirements for germination. Thus, the pattern of temperature rise in the spring has much influence on dominance hierarchy of species and therefore on community structure. Cool springs with a slow temperature rise will favor populations of *Ambrosia trifida*. Because seedlings emerge first under cool temperatures, and because they are large with actively photosynthesizing cotyledons, they can grow quickly and suppress all other species. Furthermore, because of their enormous flexibility in architecture and the ability of individuals to occupy any available space by copious branching, they can produce an almost pure stand with very high biomass production and very low species diversity (Abul-Fatih and Bazzaz 1979a). When *A. trifida* is absent or represented only very sparsely in the seed bank, *Polygonum*, whose seeds normally germinate soon after those of *Ambrosia*, dominates. Thus, dominance of field in years of cool spring is either by *Ambrosia trifida* (when its seeds are present in the site) or by *Polygonum pensylvanicum*. In contrast, a warm spring and a rapid rise in temperature will lead to an increase in the importance of *Setaria* and *Abutilon*, since under these conditions, they germinate and recruit soon after *Polygonum*. Species that germinate at warmer temperatures become increasingly more important as the time of soil disturbance is delayed. Furthermore, those species whose seeds undergo secondary dormancy become less represented in the community when soil disturbance occurs later in the growing season. As the time of soil disturbance is delayed, the cool-germinating *P. pensylvanicum* and species with secondary seed dormancy, such as *Ambrosia artemisiifolia* and *A. trifida*, become progressively less important, while *Setaria fabrei* (C_4 grass) that germinates at higher temperatures and has no secondary dormancy (Bazzaz 1984b) becomes more and more important. Thus, the combination of response to temperature, dormancy patterns, and time of soil disturbance predictably determines the general composition of this annual community in oldfields.

Growth responses to environmental factors

Plant growth characteristics that may be useful in predicting community structure include the starting capital (seed size), relative growth rate, allocation to various structures, especially leaf area ratio (LAR), and per unit leaf area photosynthetic rates. However, none of these characters leads to good predictions when taken individually. For example, maximum photosynthetic rate can be a poor predictor of the importance of a species in the community. However, maximum photosynthetic rate in conjunction with LAR can be a strong predictor. In the early-successional community, for example, even the decisively dominant *Ambrosia trifida* does not have a significantly higher photosynthetic rate per unit leaf area than many of the associated annuals (Bazzaz 1984b). However, the species can develop massive leaf areas and reallocate nitrogen quickly from shaded leaves in the dense canopy to exposed leaves at the top of the canopy. Koch *et al.* (1988)

clearly demonstrated that species with the same photosynthetic rate can grow quite differently if they shift allocation differentially among leaves, shoots, and branches.

The growth response of species of the annual community to moisture gradients resembles that of seed germination, in that the differences among species are small. All species have broad and similar responses when grown in pure stands (Fig. 12.2). However, in competitive situations, the differences among species in response to moisture become more evident. For example, the growth response of *Polygonum pensylvanicum* shifts greatly toward the wet end, whereas *Abutilon*'s response shifts somewhat toward the dry end (Pickett and Bazzaz 1978*b*) (Fig. 12.2). It is predicted, therefore, that in the field, *Polygonum* strongly expresses its dominance in wet patches, which it does. In contrast, *Ambrosia artemisiifolia* has a broad and equitable response in both pure and mixed stands. It does not shift its response when other species are present. Therefore, *Ambrosia* should be expected to be widely distributed in the field, which it is. *Abutilon* and *Amaranthus* have very similar responses in pure stands on the moisture gradient, but differ when competing with each other. *Abutilon* growth shifts slightly toward the wet end, whereas *Amaranthus* growth shifts further toward the dry end. Therefore, *Amaranthus* predominate over *Abutilon* in especially warm and dry years. Also, in dry, warm years, the C_4 plants in this community, *Amaranthus* and *Setaria*, can be more important than *Polygonum*.

In contrast to the clear differences among these annuals along light and moisture gradients, we found only small differences among them in survivorship and growth on nutrient gradients. However, among the members of this community, *Ambrosia artemisiifolia* usually has the highest concentration of nitrogen in its tissue along a broad nutrient gradient. Together with *Chenopodium*, *Ambrosia* responds to nitrogen addition by increasing uptake and foliar concentration rather than intensifying nitrogen-use efficiency. *Setaria* has the lowest levels of nitrogen in its tissue, and nitrogen concentrations in its leaves are insensitive to soil levels of nitrogen. But when these annuals are grown together in mixed stands, their responses change only slightly on the nutrient gradient. Therefore, soil nutrient level has limited predictive power in the structure of this community. Thus, we conclude that by knowing how the species in this early-successional annual community respond to their physical environment (especially temperature and soil moisture), we can make some general predictions about their relative importance in various patches in the field and in years of different climate.

Let us go back once more to the notion of competitive superiority that can greatly influence species hierarchy in communities. The only absolutely superior competitor in this system is *Ambrosia trifida* that attains dominance through early recruitment and very fast growth (Bazzaz 1984*a*). When present in an early-successional community, it decisively dominates it, producing near pure stands. Except in limited cases such as the giant ragweed (*Ambrosia trifida*) mentioned before, 'competitive superiority' is not an absolute characteristic of a species. It is a neighborhood phenomenon in that it depends on the identity of neighbors and the levels of environmental resource availability in the neighborhood. The discussion above shows that the superiority of one species over another (and perhaps its eventual replacement in succession) is relative, and may be highly influenced by the environmental circumstances in which the interaction between the two species takes place. For example, in wet locations in the field and on the moisture gradient, *Polygonum* and *Setaria* can be very important initially. Because *Setaria* is a weaker competitor, it is drastically reduced later in the season; *Amaranthus* is a very strong competitor. Similarly, under high nutrients, *Setaria* and *Chenopodium* are strong competitors, while *Amaranthus* is a weak competitor against *Setaria*. Thus, the competitive hierarchy in this community depends on the level of moisture and nutrients. It is therefore necessary to specify the environmental conditions under which competitive ability is being considered in order to predict competitive outcome. Field observations support this idea in that soil nitrogen levels greatly influence community composition (Tilman 1987, 1988). Species hierarchies also differ greatly in replicate plots differing in soil fertilizer application. Recruitment, and especially relative dominance from a uniform seed bank of annuals, greatly differ in patches with different levels of nutrient addition.

   To test if our knowledge of species response to these experimental environmental gradients and field observations are useful for predicting community structure, we quantified community composition in experimental mesocosms, developing communities from the same seed bank under a range of simulated spring conditions. Replicate mesocosms containing the same volume of soil and the same number of seeds of each of six annual species were prepared. The mesocosms were then subjected to a series of simulated springtime environmental conditions, including cool, dry, cloudy and the 1989 spring natural conditions as an unmodified reference. We also withheld soil fertilization in one treatment. The plants were allowed to grow under these different conditions in the spring for several weeks, and were then transferred until maturity to a uniform but uncontrolled summer environment. The contribution of different species to the community was

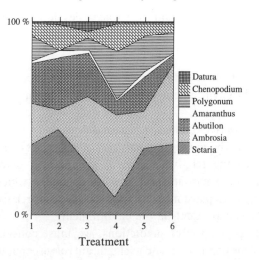

Fig. 12.3. The relative contribution to total community biomass of various early-successional annuals grown under different weather conditions in the spring (F.A. Bazzaz and P.J. Grubb unpublished data).

analyzed by canonical variants analysis. The results show that under these conditions distinct community types were produced from the same seed bank (Fig. 12.3). We also found the following:

1. *Setaria*, a C_4 grass, did well in dry conditions but poorly in cool conditions.
2. *Polygonum* did well under cool, wet conditions but did poorly under dry conditions.
3. *Abutilon* did well under dry conditions but poorly under cool conditions.
4. *Ambrosia*, the nitrogen-grabber, and *Setaria*, the efficient nitrogen-user, both did well in the low nutrient treatment.

All the results could be predicted and explained, based on our knowledge of certain aspects of the species' biology.

While there is much information about the response of plants to simple gradients, there is limited knowledge about the response of plants to multiple gradients. It is known that the response of a species on any gradient can be highly modified by the levels of other critical environmental gradients. The response surfaces of plants to critical environmental factors such as light, soil moisture, temperature, and nutrients can be even more useful in predictions, particularly for more complex successional communities made up of perennial plants with a high degree of specialization. As mentioned previously, the design and analysis of multifactorial experiments with plants is now possible and their role in developing predictive models will increase rapidly.

It is concluded that:

1. Predictions of the structure of especially short-lived communities are possible if the environment, especially soil conditions and temperature, are known.
2. Features of behavior under specific conditions that are predictive for one species may not be equally predictive for other species in the same system.
3. In order for the relative degree of 'competitive superiority' to be predictive, the environmental conditions under which competitive interactions take place must be specified, because relative competitive superiority is dependent on the environment.
4. Especially for long-term predictions, we must consider the genetic structure of the populations and its significant role in community dynamics.
5. There is no obvious reason to suspect that predictions based on physiological, demographic, and other life history features cannot be made about the structure of more complex communities, such as late-successional ecosystems.

13

From fields to forests: forest dynamics and regeneration in a changing environment

Oldfield succession, forest dynamics, and ecological theory

Many studies of secondary succession in both temperate and tropical regions have been on fields abandoned after agricultural use on previously forested lands. Oldfield succession in these areas usually leads to the development of a forest. In uncleared forests, succession proceeds from gaps of various sizes to intact forests. Pioneer herbaceous species predominate in cleared lands, while pioneer tree species predominate as early recruits in forest succession. In some tropical areas where the clearing of vegetation is incomplete and where some trees of the original forests are left standing, patchy recruitment of both herbaceous and tree pioneers can occur in the same field. In all cases in forested areas, succession, in the absence of further disturbance, ultimately leads to forest. Therefore, oldfield succession and forest succession share common features, and the species that are important in these successions share many physiological and demographic attributes.

Despite strong theoretical connections and the fact that many oldfield successions eventually result in a forest, these two areas of succession studies remain separate from one another. Our work in the Illinois fields in forest–grassland transition and later work in central New England temperate forests and the tropical rainforest in Mexico and Malaysia offers us the opportunity to discover the connections between oldfield and forest successions. It is now clear that pioneers, whether they are herbs in oldfield succession or trees in forest succession, share many characteristics with regard to seed germination, growth, and general physiology. Life history evolution of all these plants seems to have been dictated by their common presence in open, sunny habitats with greatly variable environments. Early-successional trees in temperate deciduous and moist tropical forests worldwide share many physiological and life history traits that are

239

associated with fast growth in sunny habitats (Bazzaz and Pickett 1980, Bazzaz 1990*a*).

Patterns of invasion, physiological attributes, and the resistance to invasion by various patches in oldfields were discussed in previous chapters. In this chapter, I consider gap succession that is the most common succession pattern in mature-phase forests. Most ecologists consider gap succession of critical importance to the functioning of these ecosystems, particularly in the maintenance of their biological diversity. Our approach to studying succession in forest gaps has emphasized two aspects. The first is the quantification of the changes in resource fluxes associated with the disturbances that create gaps. The second is the study of the physiological and demographic responses of various species to these changes in resource levels. We used data from these two lines of research to examine community-level attributes, such as gap partitioning among species, and their contribution to coexistence and the maintenance of species diversity. In our studies, we integrate physiological, population and community ecology. Models based on knowledge of the change in resources and the response of species to that change are used to predict forest dynamics (Carlton 1993). Similar models are being developed to study the future of forests under rapidly changing climate conditions (Pacala *et al.* 1993).

Forest succession and gap dynamics

As forests are major reservoirs of global carbon and are of great economic importance, and because they contain much of the earth's terrestrial biological diversity, their regeneration and dynamics have been extensively studied. These studies have contributed greatly to general ecological theory and have enriched our understanding of ecosystem structure and function. Models of community dynamics based on forest succession have been among the most highly developed in plant ecology. For example, the wide use of transition matrices and the modeling of succession as a Markov series (e.g. Stephens and Waggoner 1970, Horn 1971, Usher 1987, Van Hulst 1979*a,b*) have their beginnings in forest succession. The JABOWA model (Botkin *et al.* 1973) and its many modifications and extensions (e.g. FORET; Shugart *et al.* 1981) have been used to predict the effects of a variety of factors on forest growth and productivity (Shugart 1984, Botkin 1972). The new, explicitly spatial models, such as the SORTIE of Pacala *et al.* (1993), also predict forest succession.

Forest succession has also been an important vehicle for testing and expanding major ecosystem-level paradigms in ecology. In particular, response to chronic disturbances, nutrient circulation, and the links between ecosystem recovery, biomass accumulation, and nutrient retention have been tested and elaborated in forested ecosystems in different stages of succession (e.g. Woodwell 1963, Bormann and Likens 1967, 1979, Whittaker and Woodwell 1968). Forest ecosystems undergoing natural succession after experimental disturbance, such as the Hubbard Brook forest in the northeastern United States, have become touchstones for ecosystem science. It is now generally accepted that with disturbances that destroy much of the living biomass, many nutrients and much carbon are initially lost from the ecosystem to the atmosphere and to the adjacent ecosystems. However, with the development of vegetation and growth in standing biomass, nutrient export is dramatically and rapidly reduced (Vitousek and Reiners 1975). Biomass accumulation and nutrient retention are intimately connected and are the most obvious ecosystem parameters during recovery after disturbance.

Based on work in the Hubbard Brook watershed, Bormann and Likens (1979) proposed a model of biomass accumulation during succession after clearcutting. The first stage, 'reorganization,' lasts for one to two decades, during which the ecosystem loses total biomass, despite the accumulation of living biomass in the plants. The second stage, 'aggradation,' can last for up to a century, during which the system accumulates biomass to a peak. During the third stage, 'transition,' biomass declines. The fourth and final stage, 'steady state,' is when the biomass fluctuates about the mean. Interestingly, forest simulations with the JABOWA model produce similar trends of biomass accumulation (Botkin 1972). Observations by Reiners (1992) of a 20-year period of ecosystem recovery in the same watershed show that primary productivity increased exponentially first and then linearly, to about 38% of that in a mature-phase forest. The accumulation of nutrients in the aboveground biomass was significant, but different nutrients accumulated at different rates. Potassium, phosphorus, and magnesium accumulated faster, and nitrogen and calcium slower, than did biomass (Fig. 13.1). Because of these differences, ratios of nitrogen to carbon, phosphorus to carbon, nitrogen to magnesium, etc. were changed over the course of succession. These changes in stoichiometry can affect tissue chemistry, which can influence both herbivory (Mattson 1980) and litter decomposition rates (Melillo *et al.* 1982). Despite the continued changes in species composition and biomass accumulation, leaf biomass apparently stabilizes very early during forest recovery (Fig. 13.2). It is not known, however, whether gross photosynthesis by the plants is also

Forest dynamics and regeneration

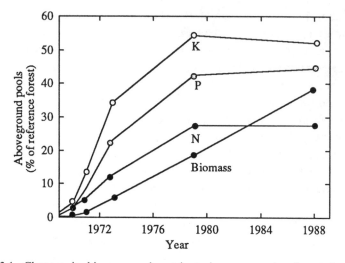

Fig. 13.1. Changes in biomass and nutrients in a recovering forest, in New Hampshire, USA (modified from Reiners 1992).

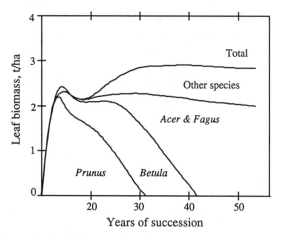

Fig. 13.2. Changes in leaf biomass of major species in succession in the Hubbard Brook ecosystem (modified from Bormann and Likens 1979).

stabilized early in recovery. Recent advances in measuring forest carbon budgets, such as eddy correlation (e.g. Wofsy *et al.* 1993), remote sensing, and scaling from leaf level measurements to ecosystem level carbon gain (e.g. Waring *et al.* 1995) promise to greatly improve the accuracy of our quantification of these budgets and our understanding of forest succession.

At the community level, much emphasis has been placed on the process of gap dynamics in forests and its role in the maintenance of species diversity (Whittaker and Levin 1977, Connell 1978, Hubbell 1979, Whitmore 1984, Brokaw 1985, Pickett and White 1985, Runkle 1985, Hubbell and Foster 1986) and the evolution of plant life history traits (Ricklefs, 1977, Denslow 1980, Bazzaz 1983, 1984*a*, 1991*b*). In this view, the forest is considered to be in a state of continuous change, and different patches in the forest are at different phases of succession. The rate of gap creation and filling is determined by several intrinsic and extrinsic factors, and can differ greatly among regions. This heterogeneity is thought to be critical in structuring distinct forest communities (Chesson and Warner 1981, Levin 1981, Strong 1983, Chesson 1985, Chesson and Huntly 1988, Diamond and Case 1986). Gaps in forests differ greatly in their size, ranging from a single tree fall to the clearing of large tracts of forests by hurricane, fire, or humans. The frequency and severity of disturbance directs the evolution of various life history designs of the species participating in the process of recovery (Levin 1976, Bazzaz 1983, 1984*b*). Long-term studies of forest dynamics, especially of the demography of major species (e.g. Piñero *et al.* 1986, Hubbell and Foster 1987, Lieberman *et al.* 1989, Clark and Clark 1991), continue to enrich our understanding of successional theory.

Gaps are created by many agents and can differ greatly in their geometry and the structure and composition of the remnant community. The identity of gap fillers is expected to be influenced by the size and severity of the gap. Branch gaps are usually filled by branch extension. Small gaps are mostly filled by the remnant community, whose growth is enhanced by resprouting and by the accelerated growth of the released, already present seedlings and saplings. Very large gaps and forest clearings, in which much of the resident vegetation is destroyed, are filled mainly by recruitment of plants from adjacent communities through seed dispersal (Fig. 13.3). In these clearings, there may be clear, repeatable patterns of recruitment in the site. Near the edge, the number of seeds arriving from the adjacent intact forest and the resulting germinated seedlings may be large and may greatly exceed the carrying capacity of the site, leading to much density-dependent mortality. Also, because of seed dispersal patterns, local recruits may be genetically closely related, setting up low diversity neighborhoods of competing relatives. Away from the edge, neighborhoods are initially less dense, more diverse, and are more likely composed of unrelated or only distantly related individuals. Despite much consideration and several tests of the Janzen–Connell hypothesis about the role of seed predators and pathogens in depressing recruitment under the mother tree (Augspurger 1983*a,b*, 1984), the degree of relatedness among neighbors in various locations in the gap,

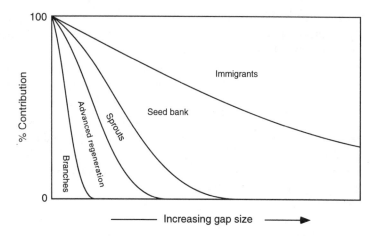

Fig. 13.3. Relationship between gap size or severity of disturbance and the relative contribution of various guilds to gap filling (modified from Bazzaz 1984*a*).

its competitive outcomes, and the evolutionary consequences of relatedness have received very little attention. An evolutionary understanding of forest succession will be greatly enhanced by studying the genetic and spatial structure of gap populations. Techniques such as DNA fingerprinting could be profitable in successional studies to explore the spatial patterns of genetic individuals and their change over successional time.

Resource fluxes in forest gaps

Forests with an intact canopy have complex vertical and horizontal environmental gradients, which are created largely by differences in the architecture of the plants and the topography of the site. In particular, light, perhaps the most critical environmental forcing function for forest productivity, has been shown to be especially heterogeneous (e.g. Chazdon 1988, Lieberman *et al.* 1989, Smith *et al.* 1989, Canham *et al.* 1990, Pearcy and Sims 1994). Irradiance, soil and air temperatures, soil nutrients and moisture, relative humidity, wind speed, and CO_2 concentrations all vary greatly in the forest and differ significantly from large gaps to the understory. The steepness of the understory-gap microenvironmental gradients is a function of disturbance frequency, disturbance size, and severity. Enormous heterogeneity in the microenvironment of gaps results from differences in gap size, orientation, and the structure of the remnant community after gap creation (Foster *et al.* 1992). This enormous complexity has made it difficult to make broad generalizations, in spite of many studies of the microenvironmental patterns in forests and gaps (see Bazzaz and Sipe 1987). Even single tree gaps can be extremely heterogeneous, especially with regard to light (Hartshorn 1978, 1980). Nutrient availability may greatly vary in a gap (Marks and Bormann 1972, Bazzaz 1983, Orians 1983, Vitousek and Denslow 1986). Therefore, we must accurately characterize this heterogeneity in order to understand patterns of recruitment and the process of gap filling.

Because of gap geometry and the position of the sun on the horizon, gaps of different sizes, shapes, and latitudes receive different amounts and quality of light (Chazdon 1986, Smith *et al.* 1989). Generally, gap centers receive more light than their edges (e.g. Denslow *et al.* 1990). Away from the tropics, the pattern of light availability also varies during the growing season (Canham *et al.* 1990). Northern parts of a gap receive more light than southern parts, and early in the season, light penetration into the intact forest extends further than late in the season. Total radiation, direct beam radiation, and daily and seasonal timing of direct beam radiation differ greatly in different locations in a gap (Fig. 13.3; Bazzaz and Wayne 1994). For example, in Harvard Forest, located in central Massachusetts, USA, northern sides of large ($375 \, m^2$) experimental gaps receive direct beam radiation for up to seven months per year, including the entire growing season. In contrast, gap centers receive only 4.5 weeks of direct beam radiation. Centers of small, single tree gaps ($75 \, m^2$) receive no direct beam radiation at all, except as sunflecks (Sipe 1990). As a result, large gaps receive two to three times as much direct beam radiation as small gaps (Fig. 13.4). Diurnal patterns of radiation also differ in different gap sizes. Large gaps experience direct beam radiation from early morning to late afternoon, while small gaps receive it only in the middle of the day. There are also great east–west phase shifts in the actual timing of direct beam radiation. Northeast parts of gaps experience direct beam radiation $2-2\frac{1}{2}$ hours later than the northwest parts. Of course, the diurnal arcs shift greatly over the course of the growing season. South sides of gaps receive diffuse radiation and are therefore lit more than the understories (Bazzaz and Wayne 1994). Thus, in temperate latitudes, there is great potential for the existence of significant microenvironmental differences within and among gaps.

In contrast, in equatorial latitudes, although there are differences in light environments at different locations, irradiance zones are more concentric (e.g. Canham *et al.* 1990, 1994). However, in some tropical environments, the afternoons are usually cloudy, and therefore, the total amount of direct-beam radiation received in the afternoon by seedlings and saplings in gaps is low relative to that received in the morning (Bazzaz 1984*b*). In this

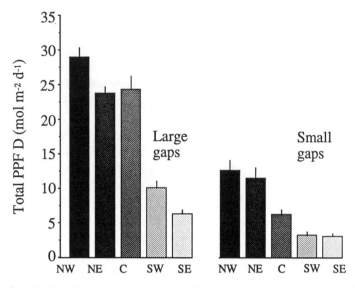

Fig. 13.4. Total daily radiation received in different locations in large (375 m²) and small (95 m²) gaps in Harvard Forest (modified from Bazzaz and Wayne 1994).

case, the amount of radiation received by seedlings at the same distance from the gap edge can differ greatly, depending on whether the seedling is located in the east or the west part of a gap. Thus, the daily integrated carbon gain may be different in different locations in a gap. Gaps are also spatially heterogeneous in terms of air and soil temperature, humidity, and wind speed associated with differences in light intensity (Sipe and Bazzaz 1995). Generally, air temperature follows direct beam radiation patterns and also varies among sites in the gap. Soil temperature usually lags behind but has the same general patterns as radiation. Near the soil surface, differences among sites in soil temperature are pronounced, but are quickly dampened with soil depth.

The gap-understory gradient in temperate forests can be summarized as follows (Sipe and Bazzaz 1995):

1. The light gradient is quite broad, spanning positions that are potentially exposed to direct irradiance for the entire growing season to positions that experience direct irradiation only as transient sunflecks.
2. The light gradient changes seasonally with solar altitude. It is not fixed in space and it fluctuates on several scales. Therefore, the gradient is truly four-dimensional.

3. The light gradient varies greatly with the prevailing weather. For example, overcast conditions dampen both spatial and diurnal variation.

4. Other environmental factors also vary both spatially and temporally, and such variation is not always in synchrony with light levels. Asynchrony or incongruence in availability of different resources is thought to be greater in gaps and other disturbed sites than in the forest understory.

Plant responses to gap environments

As described in the previous section, the interactions between light, temperature, soil and air moisture, and nutrient dynamics can be exceedingly complex. In the face of this complexity, species that occupy these sites ought to have evolved broad responses to environmental gradients. But because selection for a given trait may be offset by decreased performance by another, there are limits to selection (e.g. Antonovics 1976, Via and Lande 1985). Therefore, trade-offs are thought to exist between broad performance on a gradient and the magnitude of performance over a small range of the entire gradient. Except for the few cases mentioned before, there are apparently no forest 'super generalists' that can successfully compete along the wide gradient of resources in forest gaps (Bormann and Likens 1979, Canham and Marks 1985, Field 1991). Thus, despite their generally broad response, species can differ from each other in habitat preferences and are presumed to be capable of locating themselves in different positions along the gap-understory gradient. There is thus, the potential among species to partition the gap understory environment.

One of the most commonly used predictors of species replacement and general trends in forest dynamics is that of the differences among species in 'shade tolerance.' Intolerant (light-demanding) species are usually found early in succession, whereas shade tolerant species are recruited into the canopy later and can be assumed to successfully reproduce in their own shade. Presumably, in the absence of disturbances that drastically change the light environment in the site, shade-tolerant species can maintain their populations *in situ* for a long time.

Of course, classifications based on differences in shade tolerance implicitly include several other plant traits that are associated with light requirement. For example, with some exceptions, small seeds, high dispersal abilities, high photosynthetic and growth rate, and copious and early seed production are all important attributes of early-successional trees in both moist temperate and tropical forests (see Chapter 11). Recognizing the importance of attributes related to shade tolerance, some ecologists have grouped trees as exploitive vs. conservative, small gap vs. large gap specialists, or other dichotomous classes (e.g. Bormann and Likens 1979, Bazzaz and Pickett 1980, Denslow 1980, Lechowicz 1984, Hubbell and Foster 1986). However, within each group, species can be arranged according to finer differences in the degree of light requirement. It is agreed, for example, that within the late-successional group in the eastern deciduous forest, *Fraxinus americana* and *Quercus rubra* are generally more light-demanding than *Fagus grandifolia* and *Acer saccharum*. Refinement of these classes are possible in systems where long-term observation or copious physiological data are available. For example, in the eastern deciduous forest of North America, differences among species within each group in transition probabilities (Horn 1976), light penetration through the canopy (Canham *et al.* 1994), and physiological attributes, especially in their photosynthetic response to light, can be recognized (e.g. Bazzaz and Carlson 1982, Koike 1986, Wayne and Bazzaz 1993*a*,*b*, Bassow *et al.* 1994, Sipe and Bazzaz 1994, 1995). But because of great environmental heterogeneity, individual plasticity, and dispersal patterns, predicting the precise sequence of species occupation of gaps is difficult. Although it is of great importance to community structure, we do not know how strongly the 'principle of priority' ('first-come, first-served') operates in different ecosystems. It is tempting to ask whether low diversity ecosystems allow for more definite predictions about species sequences, as selection by some strongly limiting factors would distinctly sort out species along the successional gradient. Alternatively, it is unclear whether variable environments of such habitats have selected species which are nearly equally broad and therefore are equivalent as neighbors and are interchangeable in communities.

Partitioning the gap-understory environment among species: does it contribute to coexistence?

A central area in studies of forest dynamics has been the mechanism for the maintenance of species diversity and how that diversity changes with succession. With the rise of interest in the role played by non-equilibrium disturbance-related processes in forest structure (Hubbell 1979, Connell 1987), the 'gap-partitioning hypothesis' became more visible as a possible explanation for the maintenance of plant diversity. The hypothesis was initially proposed for tropical forests. However, there is no reason to think that it is not appropriate for other forests as well. Specialization of seedlings of different species along the microenvironmental gradient, from the understory of the intact forest to the centers of large gaps, is considered a means for habitat partitioning and niche differentiation among species. The breadth and steepness of the understory-gap microenvironmental gradient is a function of disturbance size and severity, which determine the amount and patterns of resource fluxes and the nature of the remnant community, from which recruitment into the canopy may be significant. It is not well established whether niche differentiation, leading to the occupation of different positions along these gradients in gaps, has occurred in all forests. For example, some ecologists feel that the co-occurrence of tree species in deciduous forests and mixed forests in eastern North America may reflect the present juxtaposition caused by postglacial migration patterns of species with independent evolutionary histories (e.g. Davis 1986, Foster *et al.* 1992). According to this view, coevolutionary interactions among species have little to do with the present co-occurrence. Other authors argue that forest trees, especially in more stable ecosystems, do partition environmental gradients (especially light) and therefore coexist. The gap-partitioning hypothesis (Ricklefs 1977, Bazzaz and Pickett 1980, Denslow 1980) is based on differential performance of coexisting species in gaps.

Our integrative approach to plant ecology is well suited for testing the gap-partitioning hypothesis because it requires simultaneous studies of resource fluxes and the physiological, demographic, and growth responses of the major species in a given community. We recently completed such a suite of studies in successional deciduous forests in New England. We quantified light and temperature in several locations in gaps of different sizes and in the understory and measured survivorship, growth, allocation and patterns of carbon gain of three species of maple: *Acer rubrum* (red maple), *Acer saccharum* (sugar maple), and *Acer pensylvanicum* (striped maple). The first is mid-successional, the second is a mature-phase species, and the latter is mostly an understory species (Sipe and Bazzaz 1994, 1995). We also studied the response of four birch species, *Betula populifolia* (gray birch), *Betula papyrifera* (white birch), *Betula lenta* (black birch), *Betula alleghaniensis* (yellow birch). These co-occur with the maples and are known to differ in shade tolerance (Wayne and Bazzaz 1993a). The seven species represent a spectrum, from the earliest successional, gray birch, to the most late-successional, sugar maple. Seedlings of each species were transplanted to a prescribed position in the forest. After one year of growth, gaps were created above the seedlings by felling predetermined trees. Seedlings were located in the northeast, northwest, central, and southwest parts of gaps. We wished to know if and how these species partition the gap environment.

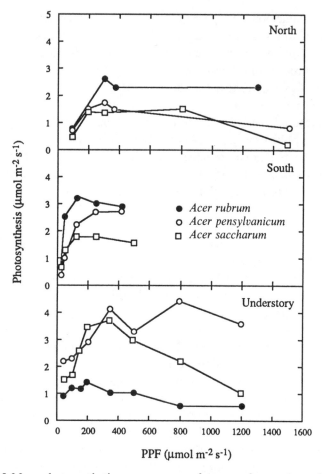

Fig. 13.5. Mean photosynthetic response curve of *Acer saccharum*, *Acer rubrum*, and *Acer pensylvanicum* to light intensity in the north and south parts of a large gap and in the forest understory (modified from Sipe and Bazzaz 1994).

Maple species showed differences in architecture that increased with time, especially in large gaps. Red and striped maple increased their branch number, leaf numbers, and total leaf area in response to gap formation much more than did sugar maple. Patterns of photosynthesis also differed for the species within and between gaps and in the understory. Red maple had the highest leaf-level photosynthetic rate, except in the south side of large gaps, where striped maple exceeded red maple (Fig. 13.5). Estimated shoot-level carbon assimilation differentiated the three maple species more than area-based photosynthetic rates, with striped maple exhibiting the greatest shoot-level rates and sugar maple the lowest. Architectural

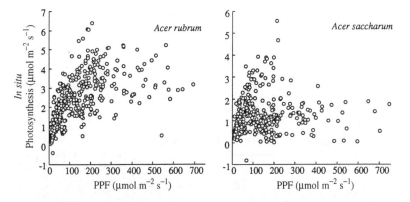

Fig. 13.6. Season-long population-level photosynthetic response of *Acer* species in a temperate forest to light in large gaps, small gaps, and the forest understory (modified from Sipe and Bazzaz 1994).

differences among the species interacted with leaf-level assimilation rates to produce differences among these species in shoot-level assimilation across the gap-understory gradient, generating some differences among them in microsite preference. Sugar maple showed a clear preference for less exposed microsites. Especially in north plots, where the radiation loads are high, this species typically closed its stomata sooner than the other two maple species when subjected to direct beam radiation, and maintained the lowest conductances of the three species throughout the day. Expressing photosynthetic rates in terms of the amount of photosynthetically active radiation (PFD) for all seedlings in all sites shows that the three species differed somewhat from each other as populations (Fig. 13.6). Population-level photosynthetic response curves reflected field situations and show which individuals were likely to perform better, survive, grow, and reach the canopy.

Among the three maple species in the study, red maple seedlings survived best, especially in the north and central parts of gaps (Fig. 13.7). Survival rates were generally high, except in exposed sites in large gaps. Differences in gap size induced greater differences among the species than were found in gap versus understory comparisons. Striped maple exhibited greater leader extension, stem height, and basal diameter than red maple and sugar maple in nearly all sites. The three species, especially red maple, experienced extensive die-back. Therefore, only striped maple showed a net increase in mean height across the entire gradient, and red and sugar maple both declined on average. Differences in growth across the gap-understory gradient paralleled patterns of shoot-level photosynthetic performance. Overall, the later-successional sugar maple showed the poorest performance.

Forest dynamics and regeneration

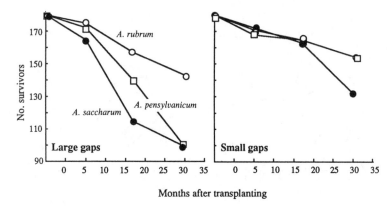

Fig. 13.7. Differential survivorship of *Acer* seedlings in relation to gap size in Harvard Forest during 35 months following seedling transplants (modified from Sipe and Bazzaz 1995).

Red maple appears to be sufficiently flexible physiologically and architecturally (Wallace and Dunn 1980, Sipe and Bazzaz 1994), to grow across a wide range of gradients as a juvenile tree. Damage to seedlings, which can be extensive in the field, greatly reduced the differences among the three species at the seedling stage, and may blur distinctions among the species with regard to gap partitioning. Without damage, however, and when considering the entire life cycle of the individual, there is little doubt that the maples are capable of partitioning the gap environment. The ranking among the species would be striped maple > red maple > sugar maple, with regard to performance across the entire microclimatic gradient in gaps. This pattern was particularly clear when the differential responses of species to the gap-understory gradient were expressed along an ordered resource gradient – a resource-response approach. Particularly revealing results can be obtained when response is plotted against several ordered relevant components of that environmental factor, such as survival against mean PPFD, optimal PPFD, sub-optimal PPFD and supra-optimal PPFD (see Sipe and Bazzaz 1995). Because a gap environment is not uniform, these three maple species behave differently in various locations.

The four species of *Betula* exhibited clearer partitioning than did the maples. For example, in understory environments, black and yellow birch showed significantly higher survivorship than the early-successional gray and white birch (Fig. 13.8). Regeneration after hurricanes of the early-successional white birch is limited to new recruitment from seed on patches of exposed soil, while yellow birch is recruited from advance regeneration, particularly from seedlings elevated above the forest floor by being close to

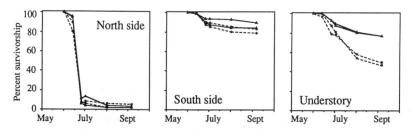

Fig. 13.8. Survivorship of four species of birch (*Betula*) in two locations in gaps and in the understory (modified from Wayne and Bazzaz 1993*a*).

the windward side of a tree that became uprooted by the hurricane (Fig. 13.9). All four birch species suffered massive mortality in the north sides of gaps. In contrast, in the south side and in the understory, the late-successional shade-tolerant species, black and yellow birch, survived better than the early-successional gray and white birch. However, all surviving individuals in the north side of the gap grew significantly larger than individuals in the south side and in the understory. Together, their survivorship and growth patterns indicate that the four birch species can co-occur by specializing in different positions along the gap-understory gradient (Wayne and Bazzaz 1993*b*). As mentioned earlier, plants in the east and west sides of gaps receive direct radiation at different hours of the day. Therefore, it is expected that plants on the west side receive maximum light in the morning when potential air, relative humidity, and CO_2 concentrations are conducive to relatively high gas exchange rates. Comparing seedlings of the early-successional gray birch and the late-successional yellow birch in experimental gaps revealed differences among them (Fig. 13.10). Taken together, the seven *Acer* and *Betula* species, which form a continuum of shade tolerance and co-occur in northern deciduous forests, show differentiation in response to light gradients (Fig. 13.11). However, it is still not known whether this partitioning is critical for the co-existence of these species.

We also tested the gap partitioning hypothesis in tropical forests in the rainforest of Southeast Asia, where despite very high tree species diversity (e.g. Richards 1952, 1995, Whitmore 1984), the family *Dipterocarpaceae* predominates and the genus *Shorea* has a large number of co-occurring species. *Shorea* is a highly speciose genus with presumably much sympatric speciation (Ashton 1969). We used several *Shorea* species to test the gap partitioning hypothesis in response to the light gradient. We measured photosynthetic rates and extension growth of a large number of seedlings of several species of *Shorea*. We quantified the light environment each seedling received by analyzing canopy photographs taken through a fish-eye lens

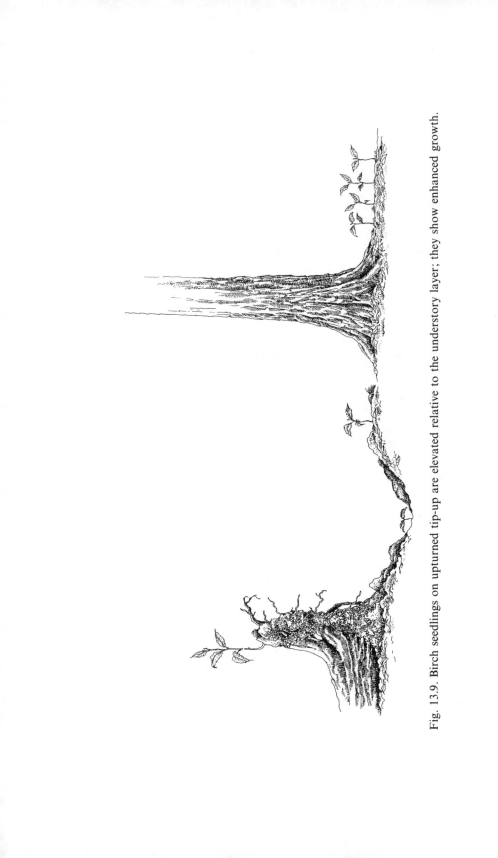

Fig. 13.9. Birch seedlings on upturned tip-up are elevated relative to the understory layer; they show enhanced growth.

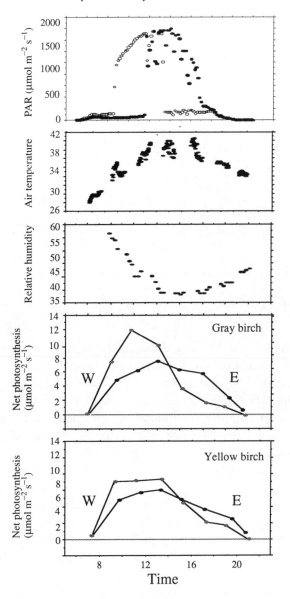

Fig. 13.10. Daily patterns of light availability, air temperature, relative humidity, and photosynthetic response of the early-successional gray birch and the late successional yellow birch located in the east and west sides of artificial gaps (modified from Wayne and Bazzaz 1993*a*).

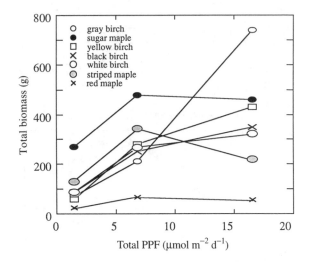

Fig. 13.11. Biomass response to increase in total photon flux of four species of birch and three species of maple forming together a gradient from the early-successional gray birch (*Betula populifolia*) to the late-successional sugar maple (*Acer saccharum*) (modified from Bazzaz and Wayne 1994).

above each seedling. We found that species response to light generally fell along a continuum from rapid potential growth and high photosynthesis to slow growth and low photosynthetic rates. There was also a clear trade-off between rapid growth in high light and persistence in low light. There were no constraints on the differentiation along the light gradient. Consequently, the species do partition the light environment (A. Moad and F.A. Bazzaz unpublished).

These studies clearly demonstrate that at the individual physiological response level, the process of gap creation generates a myriad of changes in microenvironmental resources and regulators, and elicits many interrelated responses by the plants. These studies also demonstrate that mechanistic, generalizable understanding requires close coupling among environmental, physiological, population, and community ecology. It is through this mechanistic understanding, and with appropriate modeling, that general patterns can emerge and satisfying conclusions of critical, theoretical, and applied value can be reached.

Forest regeneration following hurricane blowdown

Hurricanes and cyclones are major disturbance agents in many tropical and temperate forest ecosystems (Boose *et al.* 1994). These storms can defoliate the canopy and break or uproot trees, creating a complex array of microhabitats, including tip-up mounds, pits, boles, and crown zones (Oldeman 1978, Orians 1983, Bazzaz 1983, Walker 1991). The spatial pattern of wind damage is usually related to site exposure by wind direction, vegetation height, and species composition (Foster and Boose 1992). Microsites created by hurricanes can differ greatly in levels of environmental resources and can, therefore, have a major influence on the composition and spatial structure of the regenerating forest. Defoliation caused by hurricanes can greatly influence nutrient availability for the remaining plants and for new recruits. Wind throw creates pits and mounds (Schaetzl *et al.* 1989) and generates much microtopography in forests. Uprooting creates a variety of forest floor microsites favorable for seed germination and recruitment (Peterson and Pickett 1990) and it is known that different species can succeed in different microsites (Henry and Swan 1974, Uhl *et al.* 1981, Beatty 1984, Brandani *et al.* 1987). In many hurricanes, scattered canopy trees remain standing and others are uprooted but survive. These can be important sources of seed during the initial phases of recovery. Especially in moist tropical forests, there is usually a narrow window of recruitment time for seed dispersed into the disturbed site as seedlings and sprouts quickly form a dense layer which is hard to penetrate (Bazzaz 1984*b*).

   In order to study succession and ecosystem recovery after a hurricane, we simulated a blowdown in a mixed forest in New England. By pulling down predetermined individual trees in a predetermined direction of fall, we created a simulation of a blowdown based on patterns of damage by a natural hurricane in Harvard Forest in 1938. The simulation generated several clearly differentiated microsites in which seedling recruitment and performance were investigated. We found that these microhabitats differed greatly in light availability, CO_2 levels, soil moisture content, and nutrient relations (Fig. 13.12). Comparing variability in light and temperature in the simulated hurricane blowdown and adjacent intact forest revealed significant differences between them. Both light and temperature varied more along a transect in the blowdown than a transect in the intact forest, largely because of the presence of these distinct microhabitats in the blowdown. Also, nitrification rates and nitrate availability was greater in blowdown gaps than in the forest. At a still smaller scale, nitrification was much higher in pits than in mounds.

   The species differed in their seed dispersal patterns in these microsites. Dispersal of seeds of red maple, an important component in secondary forests in this region, was not highly influenced by microtopography.

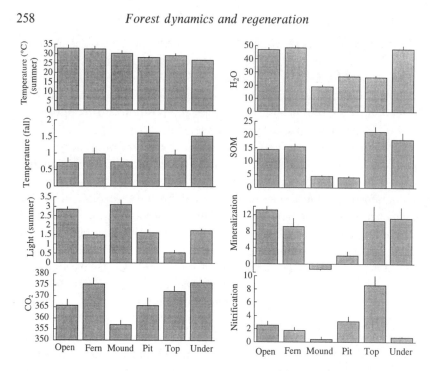

Fig. 13.12. Differences in environmental factors among microsites created by a hurricane simulation in Harvard Forest (Carlton and Bazzaz unpublished data).

However, spider webs, which are common on mounds and root plates, trap red maple seeds and may facilitate their recruitment at these sites. In contrast, red oak seeds tended to bounce off mounds and accumulate in pits. However, the great majority of them were consumed by seed predators and a high percentage of the few germinants in the pit were buried by material eroding from the mounds or suffered from waterlogging. Birch seeds also tended to collect in pits, but were not as strongly affected by microtopography as red oak seeds. Seedling mortality differed among microsites, with most seedling death occurring in the pits and least in the open (Fig. 13.13). The causes of mortality differed among sites as well (Fig. 13.14). Most mortality on pits, mounds and open sites occurred during winter and was caused by frost heaving, burial by soil and litter, and rabbit browsing. In contrast, on top and under fern microsites, most mortality occurred during the growing season and was caused mainly by resource limitation.

Growth, carbon gain, and allocation differed among sites and among species. Seedling growth was best on mound microsites and worst under ferns and other herbaceous vegetation. The early-successional white birch

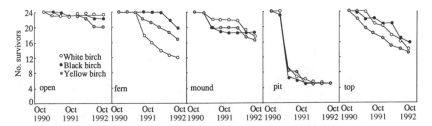

Fig. 13.13. Survivorship of white birch, black birch, and yellow birch on different microsites created by a simulated hurricane blowdown (Carlton and Bazzaz unpublished data).

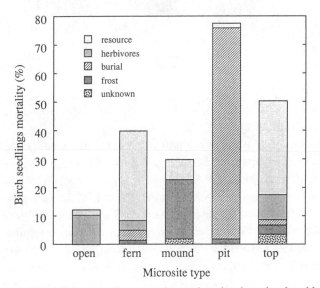

Fig. 13.14. Birch seedling mortality on various microsites in a simulated hurricane blowdown. Factors are: rabbit browsing, burial by litter and soil, frost heaving, resource limitation (Carlton and Bazzaz unpublished data).

grew faster than the late-successional yellow birch. Neither leaf nitrogen content nor water relations varied consistently by species across the different microsites. Light-saturated photosynthetic rates were relatively high in open sites and mounds and relatively low under the fern canopy. White birch had a higher light-saturated photosynthetic rate than did yellow birch, but light-limiting photosynthetic rates were similar in both species (Fig. 13.15). Elevation above the forest floor on mounds and tops gives seedlings great advantage in ascending to the canopy. After three seasons of growth, the largest seedlings were two meters tall and reached

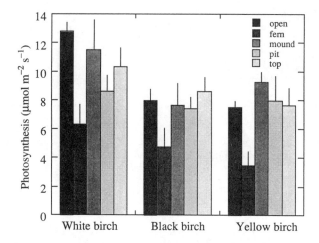

Fig. 13.15. Photosynthetic rates at saturation light conditions on various microsites of white birch, black birch, and yellow birch (Carlton and Bazzaz unpublished data).

three meters above the forest floor, well on their way to attaining reproductive maturity.

Results from this study enabled us to identify differences between white birch and yellow birch in microsite preference for regeneration. Because seedlings of white birch are unable to survive in the forest understory, regeneration of this early-successional species is limited primarily to recruitment from seed on patches of exposed soil. Many yellow birch are recruited from advance regeneration, particularly from seedlings elevated above the forest floor when canopy trees near them are uprooted. White birch and yellow birch therefore rely on two different modes of regeneration and on two different microsites created by uprooting. These two species exhibit what Peter Grubb describes as distinct regeneration niches (Grubb 1977).

There are four characteristics of tree seedling responses to their environment that add interesting complexity to the gap dynamics story:

1. Many shade-tolerant species require gaps in the canopy to reach maturity.
2. Canopy openings and sudden increases in light levels can cause severe damage or death to seedlings in the understory.
3. Many shade-tolerant species in the understory are dependent on brief periods of direct beam radiation, whereas seedlings of some pioneer species in gaps may orient their canopies to take advantage of diffuse radiation (see Ackerly and Bazzaz 1995*b*).

4. Tree seedlings, and plants in general, do not always respond positively to increased levels of light, nutrients, or water, even when these resources appear to be limiting growth.

Except for understory plants that specialize in deep shade, late-successional species are shade-tolerant, but not shade-demanding. The view that late-successional trees are able to do best in closed-canopy forests is based on the Clementsian climax concept. However, the bulk of evidence shows that the climax 'equilibrium' state is rarely attained in most forested ecosystems. Furthermore, most of the carbon-gaining portions of mature-phase trees are located in sunny environments at the top of the canopy. It is now well established that most mature phase species in temperate and tropical forests do not attain their maximal growth in shade and can benefit from gaps. In fact, many late-successional species in the rainforest even require openings in the canopy to reach reproductive maturity (e.g. Hartshorn 1980). Canham (1985) showed that *Acer saccharum*, one of the archetypal late-successional trees in the deciduous forests of eastern North America, requires two or more gaps to reach maturity. Canham (1988) contends that sugar maple, as well as *Fagus grandifolia*, the other major late-successional species in many temperate deciduous forests, are actually small gap specialists.

While canopy opening by gap creation can supply suppressed seedlings and saplings of many late-successional plants with sufficient light energy for accelerated growth and ultimate ascension into the canopy, the sudden opening of the canopy may also cause severe stress. The great increase in light levels may damage or even kill these seedlings by photoinhibition. Photoinhibition, on damage by high light levels, is usually caused by the oxidation of certain molecules critical for plant metabolism by reactive molecules, such as superoxides and hydrogen peroxides, produced because of photon excesses relative to the photon utilizing capacity of the leaves. Severe incongruency of factors, such as excess light when water is limiting, may also cause photoinhibition in plants that do not normally persist in the understory.

Sunflecks, brief periods of direct beam radiation, can contribute appreciably to carbon gain of seedlings in the forest understory and in small gaps (Pearcy *et al.* 1994). In particular, very short sunflecks are important, in that post-illumination CO_2 fixation can contribute more than expected from calculated carbon gain, based on total PPFD (Pearcy *et al.* 1994). This post-illumination CO_2 fixation occurs because of the rapid build-up of high energy metabolites, especially triose phosphate and ribulose biophosphate, whose down regulation is generally slow. Conversely, seedlings of several pioneer tree species in rain forest gaps orient their canopies toward diffuse light received from the gap rather than toward the direct sun (Ackerly and Bazzaz 1995*a,b*). This preferential orientation toward diffuse light may enhance total plant carbon gain because light saturation of photosynthesis in these species occurs at intensities below that of full sunlight. Irrespective of the actual position of these seedlings in gaps, this flexibility of canopy orientation is especially important, since spatial structure of the light environment changes as the seedlings grow through the gap. Plasticity in canopy orientation is potentially an adaptive trait for pioneers and contributes to their lifetime performance and success in a changing environment.

Results from field observations and manipulative studies suggest that plants may not always respond positively to increased resource availability. Resource augmentation as one addition or as pulses in several additions has produced mixed results. Particularly in tropical forests, the addition of nutrients has not normally increased growth (e.g. Denslow *et al.* 1990). Also, the enhancement of growth by increased light levels may be significant in some cases (e.g. Denslow *et al.* 1990) or may not (Newell *et al.* 1993), even in the same ecosystem. Multiple resource limitation and resource interactions may be stronger than is usually assumed. There are several possible explanations for the lack of response to additional resources. Competition between soil microorganisms and plants for added nitrogen may be very intense, and the microorganisms can be more successful in obtaining fertilizer inputs, especially NO_3 (Schimel *et al.* 1989). Low light levels may also severely limit the response to nutrient additions. Conversely, shortages of water or nutrients may constrain any response to added light. These studies indicate that multiple resource limitation and resource interactions may be stronger than is usually assumed. Therefore, in addition to resource abundance, we need to understand how resource levels vary over space and time and to what extent levels of different resources are congruent. Disturbance is generally thought to increase the patchiness and therefore the heterogeneity in resource levels. This may be generally true, although evidence from the Harvard Forest experimental blowdown suggests that some resources are more heterogeneous in the undisturbed forest than on a disturbed site. Disturbance events do probably tend to decrease resource congruence, as observed on the Harvard Forest experimental sites. Such incongruence may severely constrain the ability of tree seedlings to fully utilize abundant resources but may provide opportunities for compensation.

In general, studies of forest dynamics in both temperate and tropical settings have shown that the interaction between plants and their environment is extremely complex. However, as advocated throughout this book, using a resource-response approach and combining physiological and population responses can greatly inform community structure. The way in which species, early-successional as well as late-successional, respond to their environments in herbaceous communities and forests have many parallels; the successions are remarkably similar. Lessons can be extrapolated from herbaceous successional models to the forest. It is clear that causal interaction between resource availability and forest response will be further revealed through continued study of many diverse forest systems. The critical role that forests will play in the future of our planet mandates that we intensify our efforts to understand these enigmatic but precious natural ecosystems.

14

Succession and global change: will there be a shift toward more early-successional systems?

We already know that disturbances of a wide range of origins, intensities, and extents have generated successional habitats and recovering ecosystems throughout much of the earth's surface, and have influenced the evolution of both the structure and function of ecosystems. Many individual species' strategies have evolved in natural disturbance regimes and are therefore adapted to the range of circumstances that commonly occur in their habitats. In fact, because these disturbances are so prevalent, they can be considered to be components of the natural world. There is little doubt that cataclysmic events in the earth's history have led to enormous large-scale changes in ecosystems and have caused the extinction of the majority of species that ever existed. Even the eruption of a single volcano, Mt Pinatubo, has had a great impact on global oceanic and terrestrial climates. Clouds of aerosols emitted from the volcano led to a reduction in global temperature and a decline in the rate of CO_2 rise in the atmosphere, perhaps because of reduced soil respiration caused by cooling. Man-induced disturbances, such as the addition of large quantities of CO_2 and nitrogen to the atmosphere, are another kind of large-scale disturbance. In general, because of the strong feedback from the terrestrial biosphere and global climate (Woodwell et al. 1978, Mooney et al. 1987b, Sarmiento and Sundquist 1992, Woodwell and Mackenzie 1995), understanding the role of the biosphere in global change is of great importance.

Because of the recent great increase in human population and its per capita impact on the environment, concerns have arisen about human-caused disturbances. Some anthropogenic impacts can be of enormous magnitude and incalculable consequences; examples are nuclear winter (Sagan and Turco 1990), the creation of the ozone hole, and global climate change (Hougton et al. 1990, 1992). Forest decline and eventual die-back, which are occurring in many parts of the world (Schulze 1989), although assumed to be caused by several factors (Mueller-Dombois, 1992), may actually be the result, at least in part, of accelerating climate change. Nitrogen and sulfur deposition have dramatically increased in certain regions, and some of the observed forest decline in many parts of the world may be due to deposition of these chemicals (Aber *et al*. 1989, Pitelka and Raynal 1989). Nitrogen deposition is increasing worldwide and may have a great impact on ecosystem structure and function, possibly even exceeding that of rising CO_2 (Vitousek 1994). Nitrogen saturation is being approached in previously nitrogen-limited ecosystems (Aber *et al*. 1989), with the result that these systems will leak large quantities of nitrogen to the groundwater and to adjacent ecosystems. Current and future human-induced disturbances may have a very different impact on ecosystems. These disturbances are novel events that have not been experienced by the vegetation in its past history. It is therefore likely that they will have a severe impact on ecosystems. Predicting how succession proceeds after these disturbances requires a knowledge of how they might impact ecosystems differently from natural disturbances. Is the succession process after these novel disturbances governed by similar rules? In a comparative study in Harvard Forest in central Massachusetts, USA, we compared the impact of a simulated hurricane (a large tree pull-down), nitrogen-deposition, and soil warming. The first simulated a natural disturbance and the latter two simulated global change conditions generated by human activities.

In the simulated hurricane pull-down there were unexpectedly high rates of survival of damaged trees, prolific sprouting, and increase in the growth of understory plants resulting in little change in leaf area from an intact forest. Recruitment of early-successional species was unimportant and appeared to be limited to severely disturbed soil. Despite massive structural change of the forest, ecosystem-level properties changed little. For example, productivity recovered quickly, nitrogen cycling and trace gas fluxes changed little, and changes in nutrient availability were minimal. Similar situations in regard to the survival of damaged trees and their contribution to ecosystem recovery have been observed in tropical forests as well (Boucher *et al*. 1990, Whigham *et al*. 1991, Basnet 1993, Pimm *et al*. 1994). Thus despite great structural change in the forest ecosystem-level properties changed little (Foster *et al*. 1995).

In contrast, human-induced disturbances such as rapid soil warming and chronic nitrogen deposition led to great changes in the forest, despite the appearance that the vegetation was healthy. In the chronic nitrogen-addition plots there was an increase in nitrogen-leaching and a major change in trace gas emission (Aber *et al*. 1993). In the soil warming experiment there was

Table 14.1. *Impact on ecosystem properties of a simulated hurricane (natural disturbance) and a simulated chronic N-deposition and soil warming (novel, human induced disturbances) on a temperate forest*

		Human caused	
	Natural	N-deposition	Soil warming
Vegetation structure	***	●	●
Chemistry	●	*	●
Productivity	*	●	?
Soil			
Physical	●	●	***
Chemical	●	***	***
N-cycling	●	***	***
Greenhouse gas exchange	●	***	***

Unpublished data of D. Foster, J. Aber, J. Melillo, R. Bowden and F. Bazzaz.

initially a large CO_2 release from the soil (Peterjohn *et al.* 1993). These disturbances greatly altered nitrogen- and carbon-cycling; this in turn can alter ecosystem productivity (Table 14.1). There is little doubt that these systems will begin to show severe signs of decline with continued chronic disturbances. It is therefore predicted that succession and recovery will be very different from what happens after natural disturbance.

Unlike most large-scale natural events that occurred in the past, such as glaciation, these human-caused disturbances seem to occur at a greatly accelerated rate. If these events are actually taking place at this speed, most present natural ecosystems will be severely disrupted. Natural vegetation will be destabilized, since many individuals cannot function in these environments and are likely to die. The vegetation will open, and the earth will become dominated to a much greater extent by habitats undergoing succession and recovery as ecosystems begin to adjust to the new environmental conditions. Species with broad tolerances, such as weedy and early-successional species, may become dominant over a much larger area then at present. With these expectations, the study of plant succession and ecosystem recovery take on an even stronger urgency, and elucidating their general principles and rules becomes of more profound importance.

Because of the present status of world geopolitics, discussions about global scale nuclear winter have become quiescent. It is sobering to recall that under extreme nuclear winter scenarios, the earth's ecosystems can be destroyed (Ehrlich *et al.* 1984), and there will be a slow process of ecosystem repair with unknown trajectories (Bazzaz *et al.* 1985*b*). Today, there is increased interest in the response of ecosystems to the rapid climate change that is expected to occur largely due to the accumulation of the greenhouse gases (CO_2, CH_4, NO_2, CFC, etc.) in the atmosphere (e.g. Houghton and Woodwell 1989). These gases absorb infrared radiation and their increase is therefore expected to lead to a rise in global temperature. Using the current rate of increase in the concentrations of greenhouse gases in the atmosphere, scientists have made predictions about changes in temperature and precipitation patterns. Various general circulation models (GCMs) have produced different magnitudes of change (Cess *et al.* 1989), but most scientists agree that some climate change will occur. In fact, some scientists argue that an increase in global temperature is already detectable (Hanson *et al.* 1988). Others question the certainty of temperature rise and argue that other factors, such as an increase in cloud cover and reflectivity, and the negative forcing function by SO_2 in the atmosphere, may cancel out temperature increases. Others argue that CO_2 effects may be swamped by anthropogenic nitrogen and sulfur deposition (see Vitousek 1994).

If the rise in temperature, change in precipitation patterns, and increase in atmospheric CO_2 and nitrogen and sulfur deposition continue, they will have dramatic effects on ecosystems worldwide. Selective deaths of sensitive species and changes in competitive relationships may be observed in many ecosystems (Bazzaz and Fajer 1992). Also, because of the strong positive relationship between soil temperature and microbial activities, and soil respiration and CO_2 emission (e.g. Melillo *et al.* 1993, Wofsy *et al.* 1993), ecosystems such as those in northern latitudes that store large quantities of carbon may actually become major sources of carbon (Shaver *et al.* 1992, Oechel *et al.* 1994). In such cases, despite increased productivity, these ecosystems may become a major source of CO_2 released to the atmosphere. Furthermore, besides the indirect effects of CO_2 mediated through changes in the climate, CO_2 also has direct effects on primary producers in ecosystems (e.g. Strain 1987). Carbon dioxide influences photosynthetic rates, stomatal conductance, water-use efficiency, growth and allocation, and the architecture of many plant species (Strain and Cure 1985, Mooney *et al.* 1991, Woodward *et al.* 1991). It may change flowering phenology, tissue chemistry, decomposition rates, and herbivore consumption (Eamus and Jarvis 1989, Bazzaz 1990*b*). Because species of the same community may react differently both in terms of initial enhancement and later decline (Fig. 14.1), competitive interactions among plants in a community can be altered (Carlson and Bazzaz 1980, Bazzaz and McConnaughay 1992). Altered species composition may be the most critical outcome of global

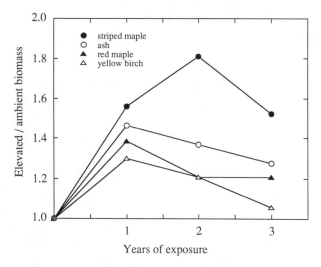

Fig. 14.1. Difference among species of a temperate forest ecosystem in enhancement and decline of plant mass ratio after three years of growth (modified from Bazzaz *et al.* 1993).

change conditions, as it may greatly influence ecosystem function and whole-ecosystem biological diversity (Bazzaz *et al.* 1995).

Feedbacks between the biosphere and the atmosphere are critical to the discussions about climate change and the future of ecosystems. Interactions between the biosphere and the atmosphere, especially the flux of CO_2, water, and isoprene, are influenced by the successional status of the impacted ecosystem. Carbon storage, net ecosystem exchange, energy budgets, water yield, and soil and plant emissions can also be influenced by successional rates and trajectories. Changes in the status of ecosystems between oxidizing (exposed, dry) and reducing (densely vegetated, wet) environments can result in changes in the identities and quantities of trace gases, such as N_2O, CO_2, and CH_4, released into the atmosphere (Mooney *et al.* 1987*b*). It is not yet known to what extent these factors might be influenced by successional patterns after the death of the present vegetation. Also, by their nature, recovering ecosystems are aggrading, and therefore must act as net sinks of carbon for some length of time. For example, a 60-year old secondary forest in New England still recovering from agricultural abandonment is currently taking up as much as 2 tons of carbon per hectare per year (Wofsy *et al.* 1993); however, it is uncertain for how long this rate of carbon accumulation will continue. In future climates, successional ecosystems can play an important role as a positive feedback on the global carbon cycle. Because of temperature increases, soil respiration may exceed net photosynthesis, even in these successional ecosystems. Under such scenarios, most ecosystems, irrespective of their successional status (and including late-successional ones), would become net CO_2 producers and a large positive feedback to atmospheric CO_2 concentration. If this is the case in future environments, one important tenet of successional theory (Odum 1969) will be violated, i.e. late-successional ecosystems can become net carbon producers.

Removal of the primary forest also alters carbon storage. It can quickly release a large quantity of carbon into the atmosphere, especially when forest clearing is followed by burning. If moisture is available, enhanced soil temperature from the initial death of vegetation may lead to a great increase in the release of soil carbon into the atmosphere as CO_2, a positive feedback on atmospheric CO_2. This situation already seems to be taking place in arctic ecosystems (Oechel *et al.*, 1994). Enhancement of plant growth by elevated CO_2 and the initial stimulation of ecosystem productivity may also increase the rate of litter accumulation, particularly if decomposition is reduced. The frequency of fire can therefore be increased. Also, the increase in non-structural carbohydrates (TNC) which has been shown to occur in plants grown in high CO_2 environments (Körner and Arnone 1992) may increase flammability. The destruction of forests can have a significant effect on the hydrological cycle as well. For example, the removal of the forest vegetation, and its replacement with grasslands and other early-successional communities, have been predicted to greatly alter the hydrological cycle in the Amazon (Sellers 1987), as a great percentage of the rainfall in the basin is recycled *in situ* in the atmosphere by local evapotranspiration (Salati *et al.* 1978). Total deforestation of the Amazon could reduce the local rainfall by 20% (Dickinson and Henderson-Sellers 1988, Lean and Warrilow 1989, Shukla *et al.* 1990).

Several investigators have developed or modified already existing models to predict the future of vegetation under climate change conditions (Shugart 1990, Prentice 1993). These predictions have produced useful insights into the future of terrestrial vegetation in a changing climate. However, in the absence of sufficient data on the response of plants to the changing climate, the limited capability of present models to accurately predict the future has been acknowledged (e.g. Shugart and Emanuel 1985). The early models have assumed, perhaps prematurely, that the direct effects of CO_2 on plant growth, reproduction, and competitive interactions are less important than the indirect climatic effects. Policy decisions should be made with caution (Fajer and Bazzaz 1992), because many of the present models are based on assumed migration rates, which are inferred from

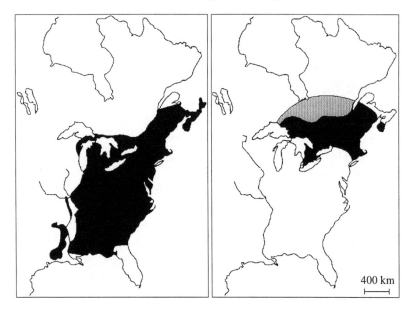

Fig. 14.2. Present and predicted future distribution for American beech (*Fagus grandifolia*) with a doubling of CO_2 in the atmosphere (modified from Davis and Zabinski 1992).

post-glacial migration patterns, and on individual species response rather than whole community response, which can be very different (Bazzaz *et al.* 1995). We now turn to a more detailed discussion of these assumptions.

Extinction, adaptation, or migration and succession

In the face of climate change, plants either adapt *in situ* to the new conditions, migrate to more suitable habitats, or become extinct. At least in the cases of extinction and migration, successional habitats will be created. Largely by analogy with post-glacial warming, several investigators have predicted a potential for large, mostly northward shifts of vegetation in the Northern Hemisphere due to species migration in response to the rising temperature (Fig. 14.2). Paleobotanical records have indicated that in the past, many plants have moved considerable distances northwards (e.g. Webb 1987) though they did not move with the same speed (Davis *et al.* 1986). The predictions of the future movement of species are based on Holocene pollen records and on inferred responses of various forest trees to air temperature, a factor which has been considered as a major determinant of large-scale plant distribution (see Woodward 1987). Using data on

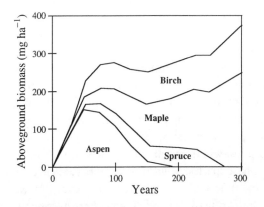

Fig. 14.3. Predicted shifts in species composition in forests in Minnesota, USA (modified from Pastor and Post 1988).

temperature rise from the GCMs, Davis and Botkin (1985), Solomon (1986), and Prentice *et al.* (1991), among others, have predicted major shifts in the eastern forests of North America as a result of the warming of the atmosphere. Similar predictions have also been made for other vegetation types in other parts of the world (e.g. Prentice 1993). On a regional scale, Pastor and Post (1988), predicted the future of forest ecosystems in Minnesota and showed the potential for change in the biomass contribution of various species over a century of temperature-driven climate change (Fig. 14.3). The predictive power of these models will undoubtedly be enhanced with the increased precision of the climate models and the incorporation of the direct effects of the CO_2 on plants and other environmental factors, such as soil properties. Many of these models considered the replacement of mature vegetation (late-successional) by other late-successional vegetation usually currently found to the south. They did not consider that any replacement in vegetation has to go through successional processes. It is, therefore, critical to remember that there is no migration that does not involve succession. Although elevated atmospheric CO_2 will increase photosynthesis and growth of plants, this enhancement differs among species and may not be long-lasting (Bazzaz *et al.* 1993). If species of the same community differ in their response to temperature extremes (e.g. Bassow *et al.* 1994) and if the enhancement ratios also differ, dominance hierarchies will be radically altered in a changing environment. It is also expected that, through altered competitive interactions and successional processes, more CO_2-responsive species could become the dominants of the system and may act as negative feedbacks on atmospheric CO_2 concentrations (e.g. Bolker *et al.* 1995).

In the northeastern United States, the work on large-scale effects and models has emphasized the migration, adaptation, and extinction of important mature-phase forest species (e.g. beech, sugar maple, and hemlock). The implicit assumption in these predictions is that species can replace each other as they march northward, and that there is enough time for this slow and more or less orderly process of migration and succession to take place. Therefore, in any given site, the mature-phase species will die, the site will open, and early-successional species, perhaps from more southern latitudes, will dominate. These will in turn be replaced by a new set of mature-phase species. Migration and replacement events are considered on a long-term time scale of hundreds of years. But with the predicted acceleration in the rate of climate change, there may be no time for current normal successional processes to take place. Additionally, in a runaway climate, there may not be enough time for natural selection to lead to the development of new, locally adapted genotypes. In this current situation there are limits to migration. Some species simply can not keep up with the rapid rate of climate change. Highly dispersed species that flower and produce seeds early in life may move long distances, but locally adapted genotypes of these species may not be able to flourish in the new habitats, especially because of the incongruence between climatic condition and soil type. Soil development, which occurs under the strong influence of both climate and vegetation, is a slow process. For example, northward migrating deciduous trees in temperate forests may find themselves on unsuitable podsolic soils that have low pH and high unincorporated organic matter and are currently occupied by conifers. Further, the extension of forest into grassland and savannahs because of the predicted increase in precipitation and water use may lead to forest growing on very different soil types. The deciduous forests of eastern North America, migrating westward, may find themselves on mollisols with much deeper, richer soils of higher pH than they presently occupy. Thus, rapid climate change will not allow rapidly migrating species to occupy suitable soil types. Unlike post-Pleistocene situations, acclimation to the new environment will be the only alternative to local extinction. Human habitations, roads, highways, etc. can also form an impervious barrier to large-scale migration. On the other hand, humans, as they are already doing, may act as agents of dispersal of selected species to new, more suitable habitats. It is also suggested that under climate change conditions exotic species can expand their current ranges. Introduction of exotic species into new habitats has significantly altered them (Mack 1986). There is little doubt that introduction of alien species whether intentional or accidental will increase. The performance of exotic species under climate change conditions is unknown. However, it has been suggested, for example, that Japanese honeysuckle (*Lonicera japonica*) and Kudzu (*Pueraria lobata*), introduced species from Japan, can greatly extend their ranges if global CO_2 concentration rise (Sasek and Strain 1990). Interactions among species will be different from what they are at present, and the outcomes of these interactions and the nature of the resulting ecosystems may be less firmly predictable. What, then, might these processes look like? What are their possible consequences?

Rapid climate change could cause some species to die locally, depending on their different sensitivities to the combination of rising temperature, changes in nutrient and water availability, rising CO_2, and the increased probability of extreme temperature events. Trees are likely to die standing. Standing death creates specific patterns of litter additions (especially coarse litter) and modifies light conditions near the forest floor in ways that are critical to the pattern of recruitment into the impacted patches. This death can also be selective, as different species in the same community may respond differently to climate change. If many of the exposed canopy trees die first, the death of some plants in lower layers in some forests can be accelerated even without climate change (e.g. Woodwell 1962, 1963). Initial recruitment will occur from advanced regeneration (seedlings and saplings) of the more resistant species and from sprouting, but if climate change continues to accelerate many of these individuals are likely to die before they reach reproductive maturity. Seed banks will be depleted by continued recruitment from, and failure of additions to, their stocks. There may be a great increase in herbaceous species. Early-successional trees that can invade the perennial herbaceous community or occupy forest gaps have physiological attributes that generally fall between those of the early-successional annuals and the late-successional trees. Therefore, as in the case of normal succession, early-successional trees are well suited to invade and eventually replace the herbaceous communities. But in the case of an accelerated rise in temperature and significant change in precipitation patterns, the course of succession may be severely modified. It is likely that late-successional, long-lived species will be less able to persist in many locations, while relatively short-lived, physiologically tolerant species may become more important components of vegetational cover, resulting in truncated successions. If climate change continues, it is very likely that an equilibrium will not be attained, and populations of late-successional species will decline greatly, particularly because they are not well-represented in seed banks. Early-successional plants will thus become strongly dominant worldwide. But if the speed of climate change declines further, and if fragmentation and man-made barriers do not seriously hamper seed dispersal, these sites can ultimately become occupied by species of more stable communities, presumably of southern origin.

While it is not possible at this time to make definitive predictions about this successional change, some possible scenarios could be considered for specific regions. Because the rate of change is so fast, plants that require a long time to become established and reproduce will be excluded from many locations. They will be present in certain patches but not as a dominant feature of large geographic areas. Another possible consequence of the continued destruction of trees, which can lead to severe soil erosion, is that herbaceous, weedy, early-successional plants, possibly including alien species, will become widespread over very large geographic areas. These species have broad responses to several environmental factors, and they possess physiological, genetic, and demographic attributes that make them successful colonizers of disturbed open sites (Bazzaz 1986, Mooney and Drake 1986). The natural cycle of regeneration will be modified such that the peaks and troughs in sizes of some populations could be greatly changed, and may be lowered to near extinction. Interestingly, one of the most prominent herbaceous genera in present-day successions, *Ambrosia*, is also well-represented in the postglaciation pollen record in the United States. Currently, several species in this genus (especially *A. artemisiifolia*, *A. elatior*, *A. trifida*, and *A. psylostachia*) are widely distributed in the temperate zone. Because of their broad morphological, physiological, and ecological variation, these species have the potential to become extremely important as early colonizers as the climate changes. But because these species depend on repeated disturbance for establishment, their populations could likely decline quickly if the climate stabilizes and will be present, as they are today, mostly in erosional patches.

Successional trajectories and climate change

Because plant resources are not independent of each other, it is reasonable to speculate that fast growing, quickly responding, opportunistic species will respond more positively than slow growing species to elevated CO_2 concentration. Available evidence suggests that fast growing early-successional trees are indeed more responsive than late-successional species (Bazzaz and Miao 1993). Evidence from herbaceous communities in England (Hunt *et al.* 1993) shows that fast growing perennials which dominate productive habitats show strong response to elevated CO_2. However, there appear to be exceptions. Mousseau and Saugier (1992) report that seedlings of the late-successional European beech, *Fagus sylvatica*, are quite responsive to elevated CO_2. The predicted widespread dominance of early- and mid-successional species under severe climate change conditions can, therefore, act as a negative feedback on the global carbon cycle as they accumulate more biomass, unless, of course, if elevated temperatures and opening of closed canopies greatly enhance soil respiration, producing a net upward flux of CO_2.

Because of great differences in the degree of disturbance and species responsiveness to elevated CO_2, succession caused by climate change can take several trajectories, depending on site characteristics, availability of seed, and chance. For example, many early-successional annuals that are dependent on annual soil disturbance may become regionally unimportant after their initial dominance. Instead, except in repeatedly disturbed habitats, it is quite likely that relatively long-lived, herbaceous, light-adapted perennial plants will become widespread in the deciduous forests of eastern North America. Of these species, the mid-successional perennial herbs *Solidago* and the grass *Andropogon* would be the most common in eastern North America and perhaps other temperate forests. Because many of these currently widely-distributed species spread clonally, reproduce sexually, and have highly dispersible seeds, they are very likely to occupy extensive areas of dead forest and remain there for several years. This, of course, does not negate the possibility that a very successful invading species that is quite responsive to climate change could dominate the local communities (e.g. Condon *et al.* 1992). Only if temperature and CO_2 concentrations are stabilized at some new level and remain so for a long time will the process of succession proceed to a regionally more or less stable state, with an increase in the representation of late-successional species. However, it is unlikely that these new ecosystems will be the same as the present ones in species composition and structure. In these systems it is likely that early- and mid-successional species will remain important for a very long time.

Litter quality and decomposition rates will change with tissue chemistry and with impacts on biogeochemical dynamics. Leaves of many plants grown in high CO_2 environments have lower nitrogen percentages and higher lignin and cellulose content relative to those grown in present-day CO_2 environments (Fig. 14.4). Experiments with leaves of high lignin content show that they decay slower than leaves with low lignin content (Melillo *et al.* 1982). Thus, leaves grown in an elevated CO_2 atmosphere may have a lower decomposition constant (K) relative to leaves grown in ambient air. It is theorized, therefore, that nutrient cycling in ecosystems, which could be enhanced initially under global warming conditions, may

Succession and global change

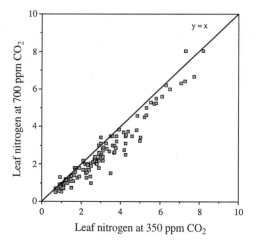

Fig. 14.4. Decline in leaf nitrogen concentration when plants are grown in a high CO_2 environment (Traw and Bazzaz unpublished data).

eventually slow down due to changes in litter chemistry. Altered nutrient cycling could then have a strong influence on successional trajectories. The resulting nutrient limitations may reduce plant CO_2 responsiveness and can influence the rate of succession. However, it should be remembered that temperature increases can increase soil microbial activities, which may make more nutrients available for roots in the soil. Furthermore, increased allocation to roots may enlarge their foraging domain and, with the possible increase in root exudates, may enhance microbial activities and increase mycorrhizal associations, which improve the availability of nutrients and the efficiency of phosphorus and nitrogen uptake by plant roots. Conversely, under these conditions, microbes encouraged by increased root exudates can vigorously compete with plant roots for available nitrogen (Fig. 14.5). It is not yet clear whether competition for nutrients will shift in favor of the plants or the soil microorganisms. Which sets of feedbacks will dominate is not yet known. Clearly, the possibilities are incredibly complex, and much study remains to be done before we can hope to get sufficient understanding from which we can draw generalizable rules about succession under conditions of significant climate change.

Plant adaptations to the new conditions

There is, of course, the possibility that plants will adapt to the global change conditions. There is genetic variation in populations' response to elevated

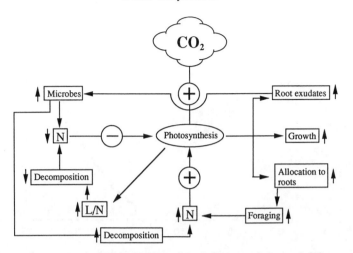

Fig. 14.5. Diagrammatic representation of the influence of elevated CO_2 on some plant and ecosystem responses.

CO_2 and temperature (Strain 1991, Curtis *et al.* 1994). This variation has been shown to be present for growth, allocation phenology of flowering, and for seed production (Garbutt and Bazzaz 1984, Bazzaz *et al.* 1995). Because of their large populations and short generation time, early-successional plants are expected to evolve CO_2-responsive genotypes faster than late-successional plants, which usually have smaller population sizes and slower generation times. However, because early-successional species are known to have broader responses and larger niche volume than late-successional plants (Chapter 8), they may be resistant to selection. This is particularly true if the populations of these species are largely made up of equally broad-niched genotypes with a high degree of flexibility. These attributes create a paradox that requires detailed investigation. On one hand, early-successional species should be subject to selection by global change conditions because of their population size and generation time; on the other hand, they resist selection because of their broad niches and high plasticity. Furthermore, the limited evidence available seems to suggest that in competitive situations (as is the case in most fields), factors other than CO_2 responsiveness may determine competitive outcomes. In other words, there may be selection largely for general competitive ability rather than for CO_2 responsiveness (Bazzaz *et al.* 1995).

The differential response of the primary producers to climate change may have consequences for other trophic levels in the ecosystem, such as herbivores (Fajer *et al.* 1992). The change in tissue carbon/nitrogen ratio

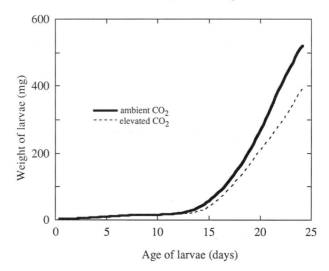

Fig. 14.6. Growth of the larvae of *Junonia* butterflies fed on *Plantago* leaves grown in ambient and enriched CO_2 environments (modified from Bazzaz and Fajer 1992).

that has been found in many plants grown in elevated CO_2 environments can have consequences for insect herbivores (e.g. Fajer *et al.* 1989, Lincoln *et al.* 1993, Lindroth *et al.* 1993). We have found that larval growth of the specialist *Junonia* butterfly is reduced when the larvae are fed leaves of *Plantago* grown in high CO_2 (Fig. 14.6). Similarly, the larvae of the generalist herbivore gypsy moth grow more slowly when fed birch leaves grown in high CO_2. There are also differences in growth depression in males and females (Fig. 14.7). It is therefore possible that herbivore populations may decrease. However, if tissue carbon/nitrogen ratios increase, herbivores must eat more leaf tissue per unit nitrogen gain. Thus, reduced nutritional quality of vegetation may elicit increased herbivory and leaf turnover (Owensby 1993), altering ecosystem level biogeochemistry and successional trajectories in as yet unknown ways.

It should be clear from the above discussion that the process of secondary succession may dominate most landscapes on earth in the future. I believe that an understanding of succession can be best obtained through comparative research in several ecosystems. I expect that in succession, life forms (vegetational physiognomy), seasonality of resource availability, and the quantity of soil resources will be more important than the identity of the species *per se*. I also believe that a resource-response perspective that considers that environmental change operates through resource modification,

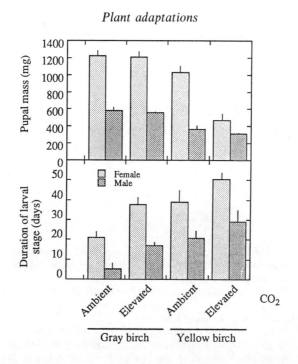

Fig. 14.7. Differential responses of male and female pupae and larvae of gypsy moths, grown on gray or yellow birch seedlings that were raised at ambient or elevated CO_2 concentrations (Traw and Bazzaz unpublished data).

and that plants respond to that change by altered growth and reproduction, is a unifying philosophy for studying successional processes, irrespective of whether they are normal or novel. Resource availability and congruence, and resource uptake, allocation, deployment, and use are the important components of this resource-response perspective. Individuals and their modules operate within the opportunites offered to, and constraints imposed by, populations, communities and ecosystems, and are rarely independent of them. Understanding succession thus requires an integrated approach. Discovering the principles of this process, the general rules that govern its patterns, speed, and trajectories, is of great importance to society. With climate change, these principles, which thus far have been of interest mainly to ecologists, will become of even greater value to society at large and particularly to decision makers responsible for generating rational policy.

References

Aarssen, L. W. 1983. Ecological combining ability and competitive combining ability in plants: Towards a general evolutionary theory of coexistence in systems of competition. *Am. Nat.* **122**: 707–31.

Aarssen, L. W. and R. Turkington. 1985. Biotic specialization between neighbouring genotypes in *Lolium perenne* and *Trifolium repens* from a permanent pasture. *J. Ecol.* **73**: 605–14.

Aber, J. D., K. J. Nadelhoffer, P. Steudler, and J. Melillo. 1989. Nitrogen saturation in northern forest ecosystems. *BioScience* **39**: 378–86.

Aber, J. D., A. Magill, R. Boone, J. M. Melillo, P. Steudler, and R. Bodwen. 1993. Plant and soil responses to chronic nitrogen additions at the Harvard Forest, Massachusetts. *Ecol. Applic.* **3**: 156–66.

Abrahamson, W. G. and H. Caswell. 1982. On the comparative allocation of biomass, energy, and nutrients in plants. *Ecology* **63**: 982–91.

Abul-Fatih, H. A. and F. A. Bazzaz. 1979a. The biology of *Ambrosia trifida* L. I. Influence of species removal on the organization of the plant community. *New Phytol.* **83**: 813–16.

Abul-Fatih, H. A. and F. A. Bazzaz. 1979b. The biology of *Ambrosia trifida* L. II. Germination, emergence, growth and survival. *New Phytol.* **83**: 817–27.

Abul-Fatih, H. A. and F. A. Bazzaz. 1980. The biology of *Ambrosia trifida* L. IV. Demography of plants and leaves. *New Phytol.* **84**: 107–11.

Abul-Fatih, H. A., F. A. Bazzaz, and R. Hunt. 1979. The biology of *Ambrosia trifida* L. III. Growth and biomass allocation. *New Phytol.* **83**: 829–38.

Ackerly, D. D. and F. A. Bazzaz. 1995a. Leaf dynamics, self-shading and carbon gain in seedlings of a tropical pioneer tree. *Oecologia* **101**: 289–98.

Ackerly, D. D. and F. A. Bazzaz. 1995b. Seedling crown orientation and interception of diffuse radiation in tropical forest gaps. *Ecology* **76**: 1134–46.

Ackerly, D. D. and M. Jasieński. 1990. Size-dependent variation of gender in high density stands of the monoecious annual, *Ambrosia artemisiifolia* (Asteraceae). *Oecologia* **82**: 474–7.

Ågren, G. I. and T. Fagerström. 1984. Limiting dissimilarity in plants: randomness prevents exclusion of species with similar competitive abilities. *Oikos* **43**: 369–75.

Alexander, H. M. 1992. Evolution of disease resistance in natural plant populations. Pages 326–44 in R. S. Fritz and E. L. Simms, eds. *Plant Resistance to Herbivores and Pathogens. Ecology, Evolution, and Genetics.* University of Chicago Press, Chicago.

Alexander, H. M. and R. D. Wulff. 1985. Experimental ecological genetics in *Plantago*. X. The effects of maternal temperature on seed and seedling characters in *P. lanceolata*. *J. Ecol.* **73**: 271–82.

Alexandré, D. Y. 1982. Aspects de la régénération naturelle en forêt dense de Côte-d'Ivoire. *Candollea* **37**: 579–88.

Allen, E. B. and M. F. Allen. 1990. The mediation of competition by mycorrhizae in successional and patchy environments. Pages 367–89 *in* J. B. Grace and D. Tilman, eds. *Perspectives on Plant Competition*. Academic Press, San Diego, CA.

Allen, M. F. 1991. *The Ecology of Mycorrhizae*. Cambridge University Press, Cambridge.

Allen, T. F. H. and T. B. Starr. 1982. *Hierarchy: Perspectives for Ecological Complexity*. University of Chicago Press, Chicago.

Alvarez-Buylla, E. R. and M. Martínez-Ramos. 1990. Seed bank versus seed rain in the regeneration of tropical pioneer trees. *Oecologia* **84**: 314–25.

Alvarez-Buylla, E. R. and R. García Barrios. 1991. Seed and forest dynamics: a theoretical framework and an example from the Neotropics. *Am. Nat.* **137**: 133–54.

Anderson, R. C., A. E. Liberta, and L. W. Dickman. 1984. Interaction of vascular plants and vesicular-arbuscular mycorrhizal fungi across a soil moisture-nutrient gradient. *Oecologia* **67**: 111–17.

Antonovics, J. 1971. The effects of a heterogeneous environment on the genetics of natural populations. *Am. Sci.* **59**: 592–9.

Antonovics, J. 1976. The nature of limits to natural selection. *Ann. Missouri Bot. Gard.* **63**: 224–47.

Antonovics, J. 1984. Genetic variation within populations. Pages 229–41 *in* R. Dirzo and J. Sarukhán, eds. *Perspectives in Plant Population Ecology*. Sinauer Associates, Sunderland, MA.

Antonovics, J. and J. Schmitt. 1986. Paternal and maternal effects on propagule size in *Anthoxanthum odoratum*. *Oecologia* **69**: 277–82.

Antonovics, J., K. Clay, and J. Schmitt. 1987. The measurement of small-scale environmental heterogeneity using clonal transplants of *Anthoxanthum odoratum* and *Danthonia spicata*. *Oecologia* **71**: 601–7.

Armesto, J. J. and S. T. A. Pickett. 1985. Experiments on disturbance in old-field plant communities: Impact on species richness and abundance. *Ecology* **66**: 230–40.

Ashmun, J. W., R. J. Thomas, and L. F. Pitelka. 1982. Translocation of photoassimilates between sister ramets in two rhizomatous forest herbs. *Ann. Bot.* **49**: 403–15.

Ashton, P. S. 1969. Speciation among tropical forest trees: some deductions in the light of recent evidence. *Biol. J. Linn. Soc.* **1**: 155–96.

Augspurger, C. K. 1983*a*. Seed dispersal of the tropical tree *Platypodium elegans*, and the escape of its seedlings from fungal pathogens. *J. Ecol.* **71**: 759–71.

Augspurger, C. K. 1983*b*. Offspring recruitment around tropical trees: changes in cohort distance with time. *Oikos* **40**: 189–96.

Augspurger, C. K. 1984. Pathogen mortality of tropical tree seedlings: experimental studies of the effects of dispersal distance, seedling density, and light conditions. *Oecologia* **61**: 211–17.

Ayala, F. 1982. The genetic structure of species. Pages 60–82 *in* R. Milkman, ed. *Perspectives in Evolution*. Sinauer Associates, Sunderland, MA.

Ayala, F. J. and C. A. Campbell. 1974. Frequency-dependent selection. *Annu. Rev. Ecol. Syst.*: 115–38.

Bacone, J., F. A. Bazzaz, and W. R. Boggess. 1976. Correlated photosynthetic responses and habitat factors of two successional tree species. *Oecologia* **23**: 63–74.

Baker, H. G. 1965. Characteristics and modes of origin of weeds. Pages 147–72 *in* H. Baker and G. Stebbins, eds. *Genetics of Colonizing Species*. Academic Press, New York.

Bakker, J. P. 1989. *Nature Management by Grazing and Cutting*. Kluwer Academic Publishers, Dordrecht.

Ballaré, C. L., R. A. Sánchez, A. L. Scopel, J. J. Casal, and C. M. Ghersa. 1987. Early detection of neighbor plants by phytochrome perception of spectral changes in reflected sunlight. *Plant, Cell Envir.* **10**: 551–7.

Bard, G. 1952. Secondary succession in the Piedmont of New Jersey. *Ecol. Monogr.* **22**: 195–215.

Barkman, J. J. 1988. New systems of plant growth forms and phenological plant types. Pages 9–44 *in* M. J. A. Werger, P. J. M. van den Aart, H. J. During, and J. T. A. Verhoeven, eds. *Plant Form and Vegetation Structure*. SPB Academic Publications, The Hague.

Barnes, P. W., W. Beyschlag, R. Ryel, S. D. Flint, and M. M. Caldwell. 1990. Plant competition for light analyzed with a multispecies canopy model. III. Influence of canopy structure in mixtures and monocultures of wheat and wild oat. *Oecologia* **82**: 560–6.

Baskin, C. and J. M. Baskin. 1988. Germination ecophysiology of herbaceous plant species in a temperate region. *Am. J. Bot.* **75**: 286–305.

Basnet, K. 1993. Recovery of a tropical rain forest after hurricane damage. *Vegetatio* **109**: 1–4.

Bassow, S. L., K. D. M. McConnaughay, and F. A. Bazzaz. 1994. The response of temperate tree seedlings grown in elevated CO_2 to extreme temperature events. *Ecol. Applic.* **4**: 593–603.

Bazzaz, F. A. 1968. Succession on abandoned fields in the Shawnee Hills, southern Illinois. *Ecology* **49**: 924–36.

Bazzaz, F. A. 1969. Succession and species distribution in relation to erosion in southern Illinois. *Trans. Ill. Acad. Sci.* **62**: 430–5.

Bazzaz, F. A. 1974. Ecophysiology of *Ambrosia artemisiifolia*: A successional dominant. *Ecology* **55**: 112–9.

Bazzaz, F. A. 1975. Plant species diversity in old-field successional ecosystems in southern Illinois. *Ecology* **56**: 485–8.

Bazzaz, F. A. 1979. The physiological ecology of plant succession. *Annu. Rev. Ecol. Syst.* **10**: 351–71.

Bazzaz, F. A. 1983. Characteristics of populations in relation to disturbance in natural and man-modified ecosystems. Pages 259–75 *in* H. A. Mooney and M. Godron, eds. *Disturbance and Ecosystems: Components of Response*. Springer-Verlag, Berlin.

Bazzaz, F. A. 1984a. Dynamics of wet tropical forests and their species strategies. Pages 233–43 *in* E. Medina, H. A. Mooney, and C. Vázquez-Yánes, eds. *Physiological Ecology of Plants of the Wet Tropics*. Dr W. Junk Publishers, The Hague.

Bazzaz, F. A. 1984b. Demographic consequences of plant physiological traits: some case studies. Pages 324–46 *in* R. Dirzo and J. Sarukhán, eds. *Perspectives on Plant Population Ecology*. Sinauer Associates, Sunderland, MA.

Bazzaz, F. A. 1986. Life history of colonizing plants: some demographic, genetic, and physiological features. Pages 96–109 *in* H. A. Mooney and J. A. Drake, eds. *Ecology of Biological Invasions of North America and Hawaii.* Springer-Verlag, New York.

Bazzaz, F. A. 1987. Experimental studies on the evolution of niche in successional plant populations: A synthesis. Pages 245–72 *in* A. J. Gray, M. J. Crawley, and P. J. Edwards, eds. *Colonization, Succession and Stability.* Blackwell Scientific Publications, Oxford.

Bazzaz, F. A. 1990*a*. Plant-plant interactions in successional environments. Pages 239–63 *in* J. B. Grace and G. D. Tilman, eds. *Perspectives on Plant Competition.* Academic Press, San Diego, CA.

Bazzaz, F. A. 1990*b*. The response of natural ecosystems to the rising global CO_2 levels. *Annu. Rev. Ecol. Syst.* **21**: 167–96.

Bazzaz, F. A. 1991*a*. Habitat selection in plants. *Am. Nat.* **137**: S116–S30.

Bazzaz, F. A. 1991*b*. Regeneration of tropical forests: physiological responses of pioneer and secondary species. Pages 91–118 *in* A. Gomez-Pompa, T. C. Whitmore, and M. Hadley, eds. *Rain Forest Regeneration and Management.* Parthenon Publishing Group, Parkridge, NJ.

Bazzaz, F. A. 1993. Scaling in biological systems: population and community perspectives. Pages 233–54 *in* J. R. Ehleringer and C. B. Field, eds. *Scaling Physiological Processes. Leaf to Globe.* Academic Press, San Diego.

Bazzaz, F. A. and D. D. Ackerly. 1992. Reproductive allocation and reproductive effort in plants. Pages 1–26 *in* M. Fenner, ed. *Seeds: The Ecology of Regeneration in Plant Communities.* CAB International, Wallingford, Oxon., UK.

Bazzaz, F. A. and R. W. Carlson. 1979. Photosynthetic contribution of flowers and seeds to reproductive effort of an annual colonizer. *New Phytol.* **82**: 223–32.

Bazzaz, F. A. and R. W. Carlson. 1982. Photosynthetic acclimation to variability in the light environment of early and late successional plants. *Oecologia* **54**: 313–16.

Bazzaz, F. A. and E. D. Fajer. 1992. Plant life in a CO_2–rich world. *Sci. Am.* **266**: 68–74.

Bazzaz, F. A. and K. Garbutt. 1988. The response of annuals in competitive neighborhoods: Effects of elevated CO_2. *Ecology* **69**: 937–46.

Bazzaz, F. A. and J. L. Harper. 1976. Relationship between plant weight and numbers in mixed populations of *Sinapsis alba* (L.) Rabenh. and *Lepidium sativum* L. *J. Appl. Ecol.* **13**: 211–16.

Bazzaz, F. A. and K. D. M. McConnaughay. 1992. Plant-plant interactions in elevated CO_2 environments. *Aust. J. Bot.* **40**: 547–63.

Bazzaz, F. A. and S. L. Miao. 1993. Successional status, seed size, and responses of tree seedlings to CO_2, light, and nutrients. *Ecology* **74**: 104–12.

Bazzaz, F. A. and S. T. A. Pickett. 1980. Physiological ecology of tropical succession: A comparative review. *Annu. Rev. Ecol. Syst.* **11**: 287–310.

Bazzaz, F. A. and E. G. Reekie. 1985. The meaning and measurement of reproductive effort in plants. Pages 373–87 *in* J. White, ed. *Studies on Plant Demography: A Festschrift for John L. Harper.* Academic Press, London.

Bazzaz, F. A. and T. W. Sipe. 1987. Physiological ecology, disturbance, and ecosystem recovery. Pages 203–27 *in* E. D. Schulze and H. Zwölfer, eds. *Potentials and Limitations of Ecosystem Analysis.* Springer-Verlag, Berlin.

Bazzaz, F. A. and S. Sultan. 1987. Ecological variation and the maintenance of plant diversity. Pages 69–93 *in* K. M. Urbanska, ed. *Differentiation Patterns in Higher Plants.* Academic Press, London.

Bazzaz, F. A. and P. Wayne. 1994. Coping with environmental heterogeneity: the physiological ecology of tree seedling regeneration across the gap-understory continuum. Pages 349–90 *in* M. M. Caldwell and R. W. Pearcy, eds. *Exploitation of Environmental Heterogeneity by Plants*: Ecophysiological Processes Above- and Belowground. Academic Press, San Diego, CA.

Bazzaz, F. A. and W. E. Williams. 1991. Atmospheric CO_2 concentrations within a mixed forest: Implications for seedling growth. *Ecology* **72**: 12–16.

Bazzaz, F. A., R. W. Carlson, and J. L. Harper. 1979. Contribution to reproductive effort by photosynthesis of flowers and fruits. *Nature* **279**: 554–5.

Bazzaz, F. A., D. A. Levin, and M. R. Schmierbach. 1982. Differential survival of genetic variants in crowded populations of *Phlox*. *J. Appl. Ecol.* **19**: 891–900.

Bazzaz, F. A., K. Garbutt, and W. E. Williams. 1985*a*. Effect of increased atmospheric carbon dioxide concentration on plant communities. Pages 155–70 *in* B. R. Strain and J. D. Cure, eds. *Direct Effects of Increasing Carbon Dioxide on Vegetation*. US Department of Energy, Washington, DC.

Bazzaz, F. A., P. Vitousek, H. A. Mooney, and R. Herrera. 1985*b*. Biological responses at the community level: background. Pages 32–6 *in* J. R. Kercher and H. A. Mooney, eds. *Research Agenda for Ecological Effects of Nuclear Winter*. National Technical Information Service, Springfield, VA.

Bazzaz, F. A., N. R. Chiariello, P. D. Coley, and L. F. Pitelka. 1987. Allocating resources to reproduction and defense. *BioScience* **37**: 58–67.

Bazzaz, F. A., K. Garbutt, E. G. Reekie, and W. E. Williams. 1989. Using growth analysis to interpret competition between a C3 and a C4 annual under ambient and elevated CO_2. *Oecologia* **79**: 223–35.

Bazzaz, F. A., J. S. Coleman, and S. R. Morse. 1990. Growth responses of seven major co-occurring tree species of the northeastern United States to elevated CO_2. *Can. J. For. Res.* **20**: 1479–84.

Bazzaz, F. A., S. L. Miao, and P. M. Wayne. 1993. CO_2-induced growth enhancements of co-occurring tree species decline at different rates. *Oecologia* **96**: 478–82.

Bazzaz, F. A., M. Jasieński, S. C. Thomas, and P. Wayne. 1995. Microevolutionary responses in experimental populations of plants to CO_2-enriched environments: parallel results from two model systems. *Proc. Natl. Acad. Sci. USA* **92**: 8161–5.

Bazzaz, F. A., S. L. Bassow, G. M. Berntson, and S. C. Thomas. 1996. Elevated CO_2 and terrestrial vegetation: implications for and beyond the global carbon budget. Pages 43–76 *in* B. Walker and W. Steffen, eds. *Global Change and Terrestrial Ecosystems*. Cambridge University Press, Cambridge.

Beattie, A. J. and D. C. Culver. 1979. Neighborhood size in Viola. *Evolution* **33**: 1226–9.

Beatty, S. W. 1984. Influence of microtopography and canopy species on spatial patterns of forest understory plants. *Ecology* **65**: 1406–19.

Begon, M., J. L. Harper, and C. Townsend. 1990. *Ecology*: Individuals, Populations, and Communities, 2nd ed. Blackwell Scientific Publications, Oxford.

Benner, B. L. and F. A. Bazzaz. 1988. Carbon and mineral element accumulation and allocation in two annual plant species in response to timing of nutrient addition. *J. Ecol.* **76**: 19–40.

Bergh, J. P. v. d. and W. G. Braakhekke. 1978. Coexistence of plant species by niche differentiation. Pages 125–38 *in* A. H. J. Freysen and J. W. Woldendorp, eds. *Structure and Functioning of Plant Populations*. North-Holland, Amsterdam.

Berntson, G. M. 1994. Modeling root architecture: Are there tradeoffs between efficiency and potential of resource acquisition? *New Phytol.* **127**: 483–93.

Berntson, G. M., E. J. Farnsworth, and F. A. Bazzaz. 1995. Allocation, within and between organs, and the dynamics of root length changes in two birch species. *Oecologia* **101**: 439–47.

Billings, W. D. 1952. The environmental complex in relation to plant growth and distribution. *Q. Rev. Biol.* **27**: 251–65.

Blom, C. W. P. M. 1978. Germination, seedling emergence and establishment of some *Plantago* species under laboratory and field conditions. *Acta Bot. Neerland.* **27**: 257–71.

Bloom, A. J., F. S. Chapin, III, and H. A. Mooney. 1985. Resource limitation in plants – an economic analogy. *Annu. Rev. Ecol. Syst.* **16**: 363–92.

Bolker, B. M., S. W. Pacala, F. A. Bazzaz, C. Canham, and S. A. Levin. 1995. Species diversity and ecosystem response to carbon dioxide fertilization: conclusions from a temperate forest model. *Global Change Biol.*

Boose, E. R., D. R. Foster, and M. Fluet. 1994. Hurricane impacts to tropical and temperate forest landscapes. *Ecol. Monogr.* **64**: 369–400.

Bormann, F. H. 1953. Factors determining the role of loblolly pine and sweetgum in early old-field succession in the Piedmont of North Carolina. *Ecol. Monogr.* **23**: 339–58.

Bormann, F. H. and G. E. Likens. 1967. Nutrient cycling. *Science* **155**: 424–9.

Bormann, F. H. and G. E. Likens. 1979. *Pattern and Process in a Forested Ecosystem*. Springer-Verlag, Berlin.

Bornkamm, R. 1984. Experimentall-ökologische untersuchungen zur Sukzession von ruderalen Pflanzengesellschaften auf unterschidlichen Böden II. Quantität und Qualität der Phytomasse. *Flora* **175**: 45–74.

Bornkamm, R. 1985. Vegetation changes in herbaceous communities. *Handb. Veget Sci* **3**: 89–109.

Bornkamm, R. 1986. Ruderal succession starting at different seasons. *Acta Soc. Bot. Pol.* **55**: 403–19.

Bornkamm, R. 1988. Mechanisms of succession on fallow lands. *Vegetatio* **77**: 95–101.

Botkin, D. B. 1972. Some ecological consequences of a computer model of forest growth. *J. Ecol.* **60**: 849–72.

Botkin, D. B., J. F. Jonak, and J. R. Wallis. 1973. Estimating the effects of carbon fertilization on forest composition by ecosystem simulation. Pages 000–00 *in* G. M. Woodwell and E. V. Pecan, eds. *Carbon and the Biosphere*. National Technical Information Service, Springfield, VA.

Boucher, D. H., J. H. Vandermeer, K. Yih, and N. Zamora. 1990. Contrasting hurricane damage in tropical rain forest and pine forest. *Ecology* **71**: 2022–4.

Boyer, J. S. 1982. Plant productivity and environment. *Science* **218**: 443–8.

Bradshaw, A. D. 1965. Evolutionary significance of phenotypic plasticity in plants. *Adv. Genet.* **13**: 115–55.

Bradshaw, A. D. 1974. Environment and phenotypic plasticity. *Brookhaven Symp. Biol.* **25**: 75–94.

Bradshaw, A. D. and M. J. Chadwick. 1980. *The Restoration of Land. The Ecology and Reclamation of Derelict and Degraded Land*. Blackwell Scientific Publications, Oxford.

Bradshaw, A. D. and K. Hardwick. 1989. Evolution and stress - genotypic and phenotypic components. *Biol. J. Linn. Soc.* **37**: 137–55.

Brandani, A., G. S. Hartshorn, and G. H. Orians. 1987. Internal heterogeneity of gaps and species richness in Costa Rican tropical rain forest. *J. Trop. Ecol.* **4**: 99–119.

Brokaw, N. V. L. 1985. Treefalls, regrowth, and community structure in tropical forests. Pages 53–69 in S. T. A. Pickett and P. S. White, eds. *The Ecology of Natural Disturbance and Patch Dynamics.* Academic Press, New York.

Brown, J. S. and D. L. Venable. 1986. Evolutionary ecology of seed-bank annuals in temporally varying environments. *Am. Nat.* **127**: 31–47.

Brown, V. K. 1985. Insect herbivores and plant succession. *Oikos* **44**: 17–22.

Brown, V. K. and T. R. E. Southwood. 1987. Secondary succession: patterns and strategies. Pages 315–37 in A. J. Gray, M. J. Crawley, and P. J. Edwards, eds. *Colonization, Succession and Stability.* Blackwell Scientific Publications, Oxford.

Burdon, J. J. 1985. *Diseases and Plant Population Biology.* Cambridge University Press, Cambridge.

Burdon, J. J., and S. R. Leather (eds.). 1990. *Pests, Pathogens and Plant Communities.* Blackwell Scientific Publications, Oxford.

Burns, R. M. and B. H. Honkala. 1990. *Silvics of North American Trees, Vol.* 2, *Hardwoods.* US Department of Agriculture, Washington, DC.

Burton, P. J. and F. A. Bazzaz. 1991. Tree seedling emergence on interactive temperature and moisture gradients and in patches of old-field vegetation. *Am. J. Bot.* **78**: 131–49.

Burton, P. J. and F. A. Bazzaz. 1995. Ecophysiological responses of tree seedlings invading different patches of old-field vegetation. *J. Ecol.* **83**: 99–112.

Caldwell, M. M. 1987. Plant architecture and resource competition. Pages 203–27 in E.-D. Schulze and H. Zwölfer, eds. *Potentials and Limitations of Ecosystem Analysis.* Springer-Verlag, New York.

Caldwell, M. M., T. J. Dean, R. S. Nowak, R. S. Dzurec, and J. H. Richards. 1983. Bunchgrass architecture, light interception, and water-use efficiency: assessment by fiber optic point quadrats and gas exchange. *Oecologia* **59**: 178–84.

Caldwell, M. M., J. H. Richards, J. H. Manwaring, and D. M. Eissenstat. 1987. Rapid shifts in phosphate acquisition show direct competition between neighboring plants. *Nature* **327**: 615–6.

Canham, C. D. 1985. Suppression and release during canopy recruitment in *Acer saccharum. Bull. Torrey Bot. Club* **112**: 134–45.

Canham, C. D. 1988. Growth and canopy architecture of shade-tolerant trees: response to canopy gaps. *Ecology* **69**: 786–95.

Canham, C. D., J. S. Denslow, W. J. Platt, J. R. Runkle, T. A. Spies, and P. S. White. 1990. Light regimes beneath closed canopies and tree-fall gaps in temperate and tropical forests. *Can. J. For. Res.* **20**: 620–31.

Canham, C. D., A. C. Finzi, S. W. Pacala, and D. H. Burbank. 1994. Causes and consequences of resource heterogeneity in forests: interspecific variation in light transmission by canopy trees. *Can. J. For. Res.* **24**: 337–49.

Canham, C. D. and P. L. Marks. 1985. The responses of woody plants to disturbance: Patterns of establishment and growth. Pages 197–217 in S. T. A. Pickett and P. S. White, eds. *The Ecology of Natural Disturbance and Patch Dynamics.* Academic Press, London.

Carlson, R. W. and F. A. Bazzaz. 1980. The effects of elevated CO_2 concentrations on growth, photosynthesis, transpiration, and water use efficiency of plants. Pages 609–23 in J. J. Singh and A. Deepak, eds. *Environmental and Climatic Impact of Coal Utilization.* Academic Press, New York.

Carlton, G. C. 1993. Effects of microsite environment on tree regeneration following disturbance. Ph. D. Thesis. Harvard University, Cambridge, MA.

Carnes, B. A. and N. A. Slade. 1982. Some comments on niche analysis in canonical space. *Ecology* **63**: 888–93.

Caswell, H. 1983. Phenotypic plasticity in life-history traits: Demographic effects and evolutionary consequences. *Am. Zool.* **23**: 35–46.

Caswell, H. 1989. *Matrix Population Models*: construction, analysis, and interpretation. Sinauer Associates, Sunderland, MA.

Causton, D. R. and J. C. Venus. 1981. *The Biometry of Plant Growth*. Edward Arnold, London.

Cess, R. D., G. L. Potter, J. P. Blanchet, G. J. Boer, S. J. Gahn, J. T. Kiehl, H. Le Treut, Z.-X. Li, X.-Z. Liang, J. F. B. Mitchell, J.-J. Morcrette, D. A. Randall, M. R. Riches, E. Roeckner, U. Schlese, A. Slingo, K. E. Taylor, M. W. M. Washington, R. T. Wetherald, and I. Yagai. 1989. Interpretation of cloud-climate feedback as produced by 14 atmospheric general circulation models. *Science* **245**: 513–16.

Chapin, F. S., III. 1993. Functional role of growth forms in ecosystem and global processes. Pages 287–312 *in* J. R. Ehleringer and C. B. Field, eds. *Scaling Physiological Processes*: Leaf to Globe. Academic Press, San Diego, CA.

Chapin, F. S., III, A. J. Bloom, C. B. Field, and R. H. Waring. 1987. Plant responses to multiple environmental factors. *BioScience* **37**: 49–57.

Chazdon, R. L. 1986. Light variation and carbon gain in rain forest understorey palms. *J. Ecol.* **74**: 995–1012.

Chazdon, R. L. 1988. Sunflecks and their importance to forest understory plants. *Adv. Ecol. Res.* **18**: 1–63.

Chesson, P. L. 1985. Coexistence of competitors in spatially and temporally varying environments: a look at the combined effects of different sorts of variability. *Theor. Pop. Biol.* **28**: 263–87.

Chesson, P. L. and N. Huntly. 1988. Community consequences of life-history traits in a variable environment. *Ann. Zool. Fenn.* **25**: 5–16.

Chesson, P. L. and R. R. Warner. 1981. Environmental variability promotes coexistence in lottery competitive systems. *Am. Nat.* **117**: 923–43.

Chiariello, N., J. C. Hickman, and H. A. Mooney. 1982. Endomycorrhizal role for interspecific transfer of phosphorus in a community of annual plants. *Science* **217**: 941–3.

Choe, H. S., C. Chu, G. Koch, J. Gorham, and H. A. Mooney. 1988. Seed weight and seed resources in relation to plant growth rate. *Oecologia* **76**: 158–9.

Clark, D. B. and D. A. Clark. 1991. The impact of physical damage on canopy tree regeneration in tropical rain forest. *J. Ecol.* **79**: 447–57.

Clausen, J., D. D. Keck, and W. M. Hiesey. 1948. *Experimental Studies on the Nature of Species. III. Environmental Responses of Climatic Races of Achillea*. Carnegie Institution of Washington Publication 581, Washington, DC.

Clay, K. 1990. The impact of parasitic and mutualistic fungi on competitive interactions among plants. Pages 391–412 *in* J. B. Grace and D. Tilman, eds. *Perspectives on Plant Competition*. Academic Press, San Diego, CA.

Clements, F. E. 1916. *Plant Succession*: An Analysis of the Development of Vegetation. Carnegie Institution of Washington Publication 520, Washington, DC.

Clough, J. M., R. S. Alberte, and J. A. Teeri. 1980. Photosynthetic adaptation of *Solanum dulcamara* L. to sun and shade environments. III. Characterization of genotypes with differing photosynthetic performance. *Oecologia* **44**: 2215.

288 *References*

Cody, M. L. 1986. Structural niches in plant communities. Pages 381–405 *in* J. Diamond and T. J. Case, eds. *Community Ecology.* Harper & Row, New York.

Coleman, J. S. and F. A. Bazzaz. 1992. Effects of CO_2 and temperature on growth and resource use of co-occurring C_3 and C_4 annuals. *Ecology* 73: 1244–59.

Colwell, R. K. and D. J. Futuyma. 1971. On the measurement of niche breadth and overlap. *Ecology* 52: 567–76.

Condon, M. A., T. W. Sasek, and B. R. Strain. 1992. Allocation patterns in two tropical vines in response to increased atmospheric carbon dioxide. *Funct. Ecol.* 6: 680–5.

Connell, J. H. 1971. On the role of natural enemies in preventing competitive exclusion in some marine animals and in rain forest trees. Pages 000–00 *in* P. J. Den Boer and G. Gradwell, eds. *Dynamics of Populations.* PUDOC, Wageningen.

Connell, J. H. 1978. Diversity in tropical rain forests and coral reefs. *Science* 199: 1302–10.

Connell, J. H. 1980. Diversity and the coevolution of competitors, or the ghost of competition past. *Oikos* 35: 131–8.

Connell, J. H. 1983. On the prevalence and relative importance of interspecific competition: Evidence from field experiments. *Am. Nat.* 122: 661–96.

Connell, J. H. 1987. Change and persistence in some marine communities. Pages 339–52 *in* A. J. Gray, M. J. Crawley, and P. J. Edwards, eds. *Colonization, Succession and Stability.* Blackwell Scientific Publications, Oxford.

Connell, J. H. 1990. Apparent versus 'real' competition in plants. Pages 9–26 *in* J. B. Grace and D. Tilman, eds. *Perspectives on Plant Competition.* Academic Press, San Diego, CA.

Connell, J. H. and R. O. Slatyer. 1977. Mechanisms of succession in natural communities and their role in community stability and organization. *Am. Nat.* 111: 1119–44.

Crabtree, R. C. and F. A. Bazzaz. 1993a. Black birch (*Betula lenta* L.) seedlings as foragers for nitrogen. *New Phytol.* 122: 617–25.

Crabtree, R. C. and F. A. Bazzaz. 1993b. Seedling response of four birch species to simulated nitrogen deposition: ammonium versus nitrate. *Ecol. Applic.* 3: 315–21.

Crawley, M. J. 1983. *Herbivory*: The Dynamics of Animal-Plant Interactions. University of California Press, Berkeley, CA.

Crawley, M. J. 1992. Seed predators and plant population dynamics. Pages 157–91 *in* M. Fenner, ed. *Seeds*: The Ecology of Regeneration in Plant Communities. CAB International, Wallingford, Oxon., UK.

Curtis, P. S., A. A. Snow, and A. S. Miller. 1994. Genotype-specific effects of elevated CO_2 on fecundity in wild radish (*Raphanus raphanistrum*). *Oecologia* 97: 100–5.

Davis, M. B. 1986. Climatic instability, time lags, and community disequilibrium. Pages 269–84 *in* J. Diamond and T. J. Case, eds. *Community Ecology.* Harper & Row, New York.

Davis, M. B. 1989. Lags in vegetation response to greenhouse warming. *Climate Change* 15: 79–82.

Davis, M. B., and D. B. Botkin. 1985. Sensitivity of cool-temperature forests and their fossil pollen record to rapid temperature change. *Quaternary Res.* 23: 327–40.

Davis, M. B., K. D. Woods, S. L. Webb, and R. P. Futyma. 1986. Dispersal versus climate: expansion of *Fagus* and *Tsuga* into the upper Great Lakes region. *Vegetatio* **67**: 93–103.

Davis, M. B. and C. Zabinski. 1992. *Changes in geographical range resulting from greenhouse warming*: effects on biodiversity in forests. Yale University Press, New Haven, CT.

Debusche, M., J. Escarre, and J. Lepart. 1982. Ornithology and plant succession in Mediterranean abandoned orchards. *Vegetatio* **48**: 255–66.

Denslow, J. S. 1980. Gap partitioning among tropical rainforest trees. *Biotropica* (*Suppl.*) **12**: 47–55.

Denslow, J. S., J. C. Schultz, P. M. Vitousek, and B. R. Strain. 1990. Growth responses of tropical shrubs to treefall gap environments. *Ecology* **71**: 165–79.

Diamond, J. and T. J. Case (eds.). 1986. *Community Ecology*. Harper & Row, New York.

Dickinson, R. E. and A. Henderson-Sellers. 1988. Modelling tropical deforestation: a study of GCM land-surface parametrizations. *Quart. J. R. Met. Soc.* **114**: 439–62.

Dinerstein, E. and C. M. Wemmer. 1988. Fruits *Rhinoceros* eat: Dispersal of *Trewia nudiflora* (Euphorbiaceae) in lowland Nepal. *Ecology* **69**: 1768–74.

Dirzo, R. 1984. Herbivory: A phytocentric overview. Pages 141–65 *in* R. Dirzo and J. Sarukhán, eds. *Perspectives in Plant Population Ecology*. Sinauer Associates, Sunderland, MA.

Dixon, R. K., S. Brown, R. A. Houghton, A. M. Solomon, M. C. Trexler, and J. Wisniewski. 1994. Carbon pools and flux of global forest ecosystems. *Science* **263**: 185–90.

Drew, A. P. and F. A. Bazzaz. 1982. Effect of night temperature on daytime stomatal conductance in early and late successional plants. *Oecologia* **54**: 76–9.

Drew, W. B. 1942. The vegetation of abandoned cropland in Cedar Creek area, Boone and Callaway Counties, Missouri. *Univ. Missouri Coll. Agric. Exp. Sta. Res. Bull.*, **344**, 52 pp.

Drury, W. H. and I. C. T. Nisbet. 1973. Succession. *J. Arnold Arbor.* **54**: 331–68.

Dueser, R. D. and H. H. Shugart Jr. 1982. Reply to comments by Van Horne & Ford and by Carnes & Slade. *Ecology* **63**: 1174–5.

Eamus, D. and P. G. Jarvis. 1989. The direct effects of increases in the global atmospheric concentrations of CO_2 on natural and commercial temperate trees and forests. *Adv. Ecol. Res.* **19**: 1–57.

Edwards, P. J. and M. P. Gillman. 1987. Herbivores and plant succession. Pages 295–314 *in* A. J. Gray, M. J. Crawley, and P. J. Edwards, eds. *Colonization, Succession and Stability*. Blackwell Scientific Publications, Oxford.

Ehleringer, J. R. and C. B. Field (eds.). 1993. *Scaling Physiological Processes*: Leaf to Globe. Academic Press, San Diego, CA.

Ehleringer, J. R., A. E. Hall, and G. D. Farquhar (eds.). 1993. *Stable Isotopes and Plant Carbon–Water Relations*. Academic Press, San Diego, CA.

Ehrlich, P., C. Sagan, D. Kennedy, and W. O. Roberts. 1984. *The Cold and the Dark*. Norton, New York.

Ellenberg, H. 1986. *Vegetation Mitteleuropas mit den Alpen*, 4th edition. Ulmer, Stuttgart.

Elton, C. S. 1927. *Animal Ecology*. Sidgwick and Jackson, London.

Ennos, R. A. 1983. Maintenance of genetic variation in plant populations. *Evol. Biol.* **16**: 129–55.

Evans, G. C. 1972. *The Quantitative Analysis of Plant Growth.* University of California Press, Berkeley, CA.

Fajer, E. D. and F. A. Bazzaz. 1992. Is carbon dioxide a 'good' greenhouse gas? Effects of increasing carbon dioxide on ecological systems. *Global Envir. Change* **2**: 301–10.

Fajer, E. D., M. D. Bowers, and F. A. Bazzaz. 1989. The effects of enriched carbon dioxide atmospheres on plant-insect herbivore interactions. *Science* **243**: 1198–200.

Fajer, E. D., M. D. Bowers, and F. A. Bazzaz. 1992. The effect of nutrients and enriched CO_2 environments on production of carbon-based allelochemicals in *Plantago*: a test of the carbon/nutrient balance hypothesis. *Am. Nat.* **140**: 707–23.

Farquhar, G. D., J. Lloyd, J. A. Taylor, L. B. Flanagan, J. P. Syvertsen, K. T. Hubick, S. C. Wong, and J. R. Ehleringer. 1993. Vegetation effects on the isotope composition of oxygen in atmospheric CO_2. *Nature* **363**: 439–43.

Farquhar, G. D., M. H. O'Leary, and J. A. Berry. 1982. On the relationship between carbon isotope discrimination and the intercellular carbon dioxide concentration in leaves. *Aust. J. Plant. Physiol.* **9**: 121–37.

Fenner, M. 1985. *Seed Ecology.* Chapman and Hall, London.

Fenner, M. (ed.) 1992. *Seeds. The Ecology of Regeneration in Plant Communities.* CAB International, Wallingford, Oxon., UK.

Ferrar, P. J., and C. B. Osmond. 1986. Nitrogen supply as a factor influencing photoinhibition and photosynthetic acclimation after transfer of shade-grown *Solanum dulcamara* to bright light. *Planta* **168**: 563–70.

Fetcher, N., S. F. Oberbauer, G. Rojas, and B. R. Strain. 1987. Efectos del régimen de luz sobre la fotosintesis y el crecimiento en plántulas de árboles de un bosque lluvioso tropical de Costa Rica. *Rev. Biol. Trop.* **35 (Supl.)**: 97–110.

Field, C. B. 1983. Allocating leaf nitrogen for the maximization of carbon gain: Leaf age as a control on the allocation program. *Oecologia* **56**: 341–7.

Field, C. B. 1991. Ecological scaling of carbon gain to stress and resource availability. Pages 35–65 *in* H. A. Mooney, W. E. Winner and E. J. Pell, eds. *Response of Plants to Multiple Stresses.* Academic Press, San Diego, CA.

Finegan, B. 1984. Forest succession. *Nature* **312**: 109–14.

Fitter, A. 1987. *Environmental Physiology of Plants.* Academic Press, London.

Foin, T. C. and S. K. Jain. 1977. Ecosystems analysis and population biology: lessons for the development of community ecology. *BioScience* **27**: 532–8.

Foster, D. R. 1988. Species and stand response to catastrophic wind in central New England, USA *J. Ecol.* **76**: 135–51.

Foster, D. R. and E. R. Boose. 1992. Patterns of forest damage resulting from catastrophic wind in central New England, USA. *J. Ecol.* **80**: 79–98.

Foster, D. R., T. Zebryk, P. Schoonmaker, and A. Lezberg. 1992. Post-settlement history of human land-use and vegetation dynamics of a *Tsuga canadensis* (hemlock) woodlot in central New England. *J. Ecol.* **80**: 773–86.

Foster, D. R., J. D. Aber, J. M. Mellilo, R. D. Bowden, and F. A. Bazzaz. 1995. Temperate forest response to natural catastrophic disturbance and chronic anthropogenic stress. *BioScience* (in press).

Foster, S. A. 1986. On the adaptive value of large seeds for tropical moist forest trees: a review and synthesis. *Bot. Rev.* **53**: 260–99.

Franco, M. 1986. The influence of neighbors on the growth of modular organisms with an example from trees. *Phil. Tran. R. Soc. Lond. B* **313**: 209–25.

Franz, E. H. and F. A. Bazzaz. 1977. Simulation of vegetation response to modified hydrologic regimes: A probabilistic model based on niche differentiation in a floodplain forest. *Ecology* **58**: 176–83.

Fritz, R. S. and E. L. Simms (eds.). 1992. *Plant Resistance to Herbivores and Pathogens. Ecology, Evolution, and Genetics*. University of Chicago Press, Chicago.

Garbutt, K. and F. A. Bazzaz. 1984. The effects of elevated CO_2 on plants III. Flower, fruit and seed production and abortion. *New Phytol.* **98**: 433–46.

Garbutt, K. and F. A. Bazzaz. 1987. Population niche structure. Differential response of *Abutilon theophrasti* progeny to resource gradients. *Oecologia* **72**: 291–6.

Garbutt, K. and A. R. Zangerl. 1983. Application of genotype-environment interaction analysis to niche quantification. *Ecology* **64**: 1292–6.

Garbutt, K., W. E. Williams, and F. A. Bazzaz. 1990. Analysis of the differential response of five annuals to elevated CO_2 during growth. *Ecology* **71**: 1185–94.

Gates, D. M. 1965. Energy, plants, and ecology. *Ecology* **46**: 1–13.

Gaudet, C. L. and P. A. Keddy. 1988. A comparative approach to predicting competitive ability from plant traits. *Nature* **334**: 242–3.

Geber, M. A. 1990. The cost of meristem limitation in *Polygonum arenastrum*: negative genetic correlations between fecundity and growth. *Evolution* **44**: 799–819.

Geiger, R. 1965. *The Climate Near the Ground*. Translation of German 4th edition. Harvard University Press, Cambridge, MA.

Gill, D. S. and P. L. Marks. 1991. Tree and shrub seedling colonization of old fields in Central New York. *Ecol. Monogr.* **61**: 183–205.

Givinish, T. J. 1988. Adaptations to sun and shade: a whole plant perspective. *Aust. J. Plant Physiol.* **15**: 63–92.

Gleason, H. A. 1926. The individualistic concept of plant association. *Bull. Torrey Bot. Club* **53**: 7–26.

Godfray, H. C. J. 1985. The absolute abundance of leaf miners on plants of different successional stages. *Oikos* **45**: 17–25.

Goldberg, D. E. 1987. Neighborhood competition in an old-field plant community. *Ecology* **68**: 1211–23.

Goldberg, D. E. 1990. Components of resource competition in plant communities. Pages 27–49 *in* J. B. Grace and D. Tilman, eds. *Perspectives on Plant Competition*. Academic Press, San Diego, CA.

Goldberg, D. E. and K. L. Gross. 1988. Disturbance regimes of midsuccessional old fields. *Ecology* **69**: 1677–88.

Goldberg, D. E. and P. A. Werner. 1983. Equivalence of competitors in plant communities: A null hypothesis and a field experimental approach. *Am. J. Bot.* **70**: 1098–104.

Goloff, A. A., Jr. and F. A. Bazzaz. 1975. A germination model for natural seed populations. *J. Theor. Biol.* **52**: 259–83.

Gorham, E., P. M. Vitousek, and W. A. Reiners. 1979. The regulation of chemical budgets over the course of terrestrial ecosystem succession. *Annu. Rev. Ecol. Syst.* **10**: 53–88.

Grace, J. B. 1990. On the relationship between plant traits and competitive ability. Pages 51–65 *in* J. B. Grace and D. Tilman, eds. *Perspectives on Plant Competition*. Academic Press, San Diego, CA.

Grace, J. B., and D. Tilman (eds.). 1990. *Perspectives on Plant Competition*. Academic Press, San Diego, CA.

Gray, A. J. 1987. Genetic change during succession in plants. Pages 273–93 *in* A. J. Gray, M. J. Crawley, and P. J. Edwards, eds. *Colonization, Succession and Stability*. Blackwell Scientific Publications, Oxford.

Grime, J. P. 1977. Evidence for the existence of three primary strategies in plants and its relevance to ecological and evolutionary theory. *Am. Nat.* **11**: 1169–94.

Grime, J. P. 1979. *Plant Strategies and Vegetation Processes*. John Wiley, Chichester, UK.

Grime, J. P. and D. W. Jeffrey. 1965. Seedling establishment in vertical gradients of sunlight. *J. Ecol.* **53**: 621–42.

Grinnell, J. 1928. Presence and absence of animals. *Univ. California Chronicle* **30**: 429–50.

Gross, K. L. 1984. Effects of seed size and growth form on seedling establishment of six monocarpic perennial plants. *J. Ecol.* **72**: 369–87.

Gross, K. L. 1980. Colonization by *Verbascum thapsus* (mullein) of an old-field in Michigan: Experiments on the effects of vegetation. *J. Ecol.* **68**: 919–27.

Grubb, P. J. 1977. The maintenance of species richness in plant communities: the importance of the regeneration niche. *Biol. Rev.* **52**: 107–45.

Hall, F. G., D. B. Botkin, D. E. Strebel, K. D. Woods, and S. J. Goetz. 1991. Large-scale patterns of forest succession as determined by remote sensing. *Ecology* **72**: 628–40.

Hamrick, J. L., Y. B. Linhart, and J. B. Mitton. 1979. Relationships between life history characteristics and electrophoretically detectable genetic variation in plants. *Annu. Rev. Ecol. Syst.* **10**: 173–200.

Hanson, J., I. Fung, A. Lacias, D. Rind, G. Russell, S. Lebedeff, R. Rudy, and P. Stone. 1988. Global climate change as forecast by the GISS 3-D model. *J. Geophys. Res.* **93**: 9341–64.

Hanzawa, F. M., A. J. Beattie, and D. C. Culver. 1988. Directed dispersal: demographic analysis of an ant-seed mutualism. *Am. Nat.* **131**: 1–13.

Harcombe, P. A. 1977. The influence of fertilization on some aspects of succession in a humid tropical forest. *Ecology* 58: 1375–83.

Harper, J. L. 1967. A Darwinian approach to plant ecology. *J. Ecol.* **55**: 247–70.

Harper, J. L. 1969. The role of predation in vegetational diversity. *Brookhaven Symp. Biol.* **22**: 48–62.

Harper, J. L. 1977. *The Population Biology of Plants*. Academic Press, London.

Harper, J. L. 1985. Modules, branches and the capture of resources. Pages 1–34 *in* J. B. C. Jackson, L. W. Buss, and R. E. Cook, eds. *Population Biology and Evolution of Clonal Organisms*. Yale University Press, New Haven, CT.

Harper, J. L. 1988. An apophasis of plant population biology. Pages 435–52 *in* A. J. Davy, M. J. Hutchings, and A. R. Watkinson, eds. *Plant Population Ecology*. Blackwell Scientific Publications, Oxford.

Harper, J. L. 1989. Canopies as populations. Pages 105–28 *in* G. Russell, B. Marshall, and P. G. Jarvis, eds. *Plant Canopies: Their Growth, Form and Function*. Cambridge University Press, Cambridge.

Harper, J. L., P. H. Lovell, and K. G. Moore. 1970. The shapes and sizes of seeds. *Annu. Rev. Ecol. Syst.* **1**: 327–56.

Harrison, J. S. and P. A. Werner. 1984. Colonization by oak seedlings into a heterogeneous successional habitat. *Can. J. Bot.* **62**: 559–63.

Hartgerink, A. P. 1981. Effects of Some Events Experienced by Seedlings in Competition. Ph. D. Thesis. University of Illinois, Urbana.

Hartgerink, A. P. and F. A. Bazzaz. 1984. Seedling-scale environmental heterogeneity influences individual fitness and population structure. *Ecology* **65**: 198–206.

Hartnett, D. C. 1990. Size-dependent allocation to sexual and vegetative reproduction in four clonal composites. *Oecologia* **84**: 254–9.

Hartnett, D. C. and F. A. Bazzaz. 1983. Physiological integration among intraclonal ramets in *Solidago canadensis*. *Ecology* **64**: 779–88.

Hartnett, D. C. and F. A. Bazzaz. 1985a. The genet and ramet population dynamics of *Solidago canadensis* in an abandoned field. *J. Ecol.* **73**: 407–13.

Hartnett, D. C. and F. A. Bazzaz. 1985b. The integration of neighbourhood effects by clonal genets of *Solidago canadensis*. *J. Ecol.* **73**: 415–27.

Hartnett, D. C. and F. A. Bazzaz. 1985c. The regulation of leaf, ramet and genet densities in experimental populations of the rhizomatous perennial *Solidago canadensis*. *J. Ecol.* **73**: 429–43.

Hartnett, D. C., B. B. Hartnett, and F. A. Bazzaz. 1987. Persistence of *Ambrosia trifida* populations in old fields and responses to successional changes. *Am. J. Bot.* **74**: 1239–48.

Hartshorn, G. 1978. Tree falls and tropical forest dynamics. Pages 617–38 *in* P. B. Tomlinson and M. H. Zimmerman, eds. *Tropical Trees as Living Systems*. Cambridge University Press, Cambridge.

Hartshorn, G. S. 1980. Neotropical forest dynamics. *Biotropica*. **12**: 23–30.

Hayashi, I. 1977. Secondary succession of herbaceous communities in Japan. *Jap. J. Ecol.* **27**: 191–200.

Hedrick, P. W. 1986. Genetic polymorphism in heterogeneous environments: a decade later. *Annu. Rev. Ecol. Syst.* **17**: 535–66.

Hendrix, S. D., V. K. Brown, and H. Dingle. 1988a. Arthropod guild structure during early old field succession in a New and Old World site. *J. Anim. Ecol.* **57**: 1053–66.

Hendrix, S. D., V. K. Brown, and A. C. Gange. 1988b. Effects of insect herbivory on early plant succession: comparison of an English (UK) site and an American site. *Biol. J. Linn. Soc.* **35**: 205–16.

Henry, J. D. and J. M. A. Swan. 1974. Reconstructing forest history from live and dead plant material – an approach to the study of forest succession in southwest New Hampshire. *Ecology* **55**: 772–83.

Herrera, C. M. and P. Jordano. 1981. *Prunus mahaleb* and birds: The high-efficiency seed dispersal system of a temperate fruiting tree. *Ecol. Monogr.* **51**: 203–18.

Hirose, T. and M. J. A. Werger. 1987. Maximizing daily canopy photosynthesis with respect to the leaf nitrogen allocation pattern in the canopy. *Oecologia* **72**: 520–6.

Holbrook, N. M. and F. E. Putz. 1989. Influence of neighbors on tree form: effects of lateral shade and prevention of sway on the allometry of *Liquidambar styraciflua* (sweet gum). *Am. J. Bot.* **76**: 1740–9.

Horn, H. S. 1971. *The Adaptive Geometry of Trees*. Princeton University Press, Princeton, NJ.

Horn, H. S. 1975. Markovian processes of forest succession. Pages 196–211 *in* M. L. Cody and J. M. Diamond, eds. *Ecology and Evolution of Communities*. Belknap Press, Cambridge, MA.

Horn, H. S. 1976. Succession. Pages 187–204 *in* R. M. May, ed. *Theoretical Ecology. Principles and Applications*. W. B. Saunders, Philadelphia, PA.

Horvitz, C. C. and D. W. Schemske. 1988. Demographic cost of reproduction in a neotropical herb: an experimental approach. *Ecology* **69**: 1741–5.

Houghton, J. T., G. J. Jenkins, and J. J. Ephraums (eds.). 1990. *Climate Change*: The IPCC Scientific Assessment. Cambridge University Press, Cambridge.

Houghton, J. T., B. A. Callander and S. K. Varney (eds.). 1992. *Climate Change 1992*: The Supplementary Report to The IPCC Scientific Assessment. Cambridge University Press, Cambridge.

Houghton, R. A. and G. M. Woodwell. 1989. Global climatic change. *Sci. Am.* **260**: 36–44.

Hubbell, S. P. 1979. Tree dispersion, abundance, and diversity in a tropical dry forest. *Science* **203**: 1299–309.

Hubbell, S. P. and R. B. Foster. 1986. Canopy gaps and the dynamics of neotropical forest. Pages 73–96 *in* M. J. Crawley, ed. *Plant Ecology*. Blackwell Scientific Publications, Oxford.

Hubbell, S. P. and R. B. Foster. 1987. The spatial context of regeneration in a Neotropical forest. Pages 395–412 *in* A. J. Gray, M. J. Crawley, and P. J. Edwards, eds. *Colonization, Succession and Stability*. Blackwell Scientific Publications, Oxford.

Hunt, R. 1982. *Plant Growth Curves*: An Introduction to the Functional Approach to Plant Growth Analysis. Edward Arnold, London.

Hunt, R. 1990. *Basic Growth Analysis*. Unwin Hyman, London.

Hunt, R., D. W. Hand, M. A. Hannah, and A. M. Neal. 1993. Further responses to CO2 enrichment in British herbaceous species. *Funct. Ecol.* **7**: 661–8.

Huston, M. 1979. A general hypothesis of species diversity. *Am. Nat.* **113**: 81–101.

Huston, M. and T. Smith. 1987. Plant succession: life history and competition. *Am. Nat.* **130**: 168–98.

Huston, M. A. 1994. *Biological Diversity. The Coexistence of Species on Changing Landscapes*. Cambridge University Press, Cambridge.

Hutchings, M. J. 1988. Differential foraging for resources and structural plasticity in plants. *Trends Ecol. Evol.* **3**: 200–4.

Hutchinson, G. 1957. Concluding remarks. *Cold Spring Harbor Symp. Quant. Biol.* **22**: 415–27.

Inouye, R. S., N. J. Huntly, D. Tilman, J. R. Tester, M. Stillwell, and K. C. Zinnel. 1987. Old-field succession on a Minnesota sand plain. *Ecology* **68**: 12–26.

Jacquard, P. 1968. Manifestacion et nature des relations sociales chez les végétaux superieurs. *Oecol. Plant.* **3**: 137–68.

Jain, S. 1979. Adaptive strategies: polymorphism, plasticity, and homeostasis. Pages 160–87 *in* O. T. Solbrig, S. Jain, G. B. Johnson, and P. H. Raven, eds. *Topics in Plant Population Biology*. Columbia University Press, New York.

Jain, S. K. 1990. Variation and selection in plant populations. Pages 199–230 *in* K. Wohrmann and S. K. Jain, eds. *Population Biology*: Ecological and Evolutionary Viewpoints. Springer-Verlag, Berlin.

Janos, D. P. 1980. Mycorrhizae influence tropical succession. *Biotropica* **12** (Suppl.): 56–64.

Janzen, D. H. 1970. Herbivores and the number of tree species in tropical forests. *Am. Nat.* **104**: 501–29.

Janzen, D. H. 1981. Patterns of herbivory in a tropical deciduous forest. *Biotropica* **13**: 271–82.

Jones, M. and J. L. Harper. 1987. The influence of neighbors on the growth of trees. I. The demography of buds in *Betula pendula*. *Proc. R. Soc. Lond. B* **232**: 1–18.

Jurik, T. W. 1985. Differential costs of sexual and vegetative reproduction in wild strawberry populations. *Oecologia* **66**: 294–403.

Keever, C. 1950. Causes of succession in old fields of the Piedmont, North Carolina. *Ecol. Monogr.* **20**: 229–50.

Kirkpatrick, B. L. and F. A. Bazzaz. 1979. Influence of certain fungi on seed germination and seedling survival of four colonizing annuals. *J. Appl. Ecol.* **16**: 515–27.

Koch, G. W., E. D. Schulze, F. Percival, H. A. Mooney, and C. Chu. 1988. The nitrogen balance of Raphanus sativus x raphanistrum plants. II. Growth, nitrogen redistribution and photosynthesis under NO_3–deprivation. *Plant, Cell Envir.* **11**: 755–67.

Koide, R. T., D. L. Shumway, and S. A. Mabon. 1994. Mycorrhizal fungi and reproduction of field populations of *Abutilon theophrasti* Medic. (Malvaceae). *New Phytol.* **126**: 123–30.

Koike, T. 1986. Photosynthetic responses to light intensity of deciduous broad-leaved tree seedlings raised under various artificial shade. *Envir. Conserv. Biol.* **24**: 51–8.

Koike, T. 1988. Leaf structure and photosynthetic performance as related to the forest succession of deciduous broad-leaved trees. *Plant Species Biol.* **3**: 77–87.

Kolasa, J. and S. T. A. Pickett (eds.). 1991. *Ecological Heterogeneity.* Springer-Verlag, New York.

Kolasa, J. and C. D. Rollo. 1991. Introduction: The heterogeneity of heterogeneity: A glossary. Pages 1–23 *in* J. Kolasa and S. T. A. Pickett, eds. *Ecological Heterogeneity.* Springer-Verlag, New York.

Körner, C. and J. A. Arnone, III. 1992. Responses to elevated carbon dioxide in artificial tropical ecosystems. *Science* **257**: 1672–5.

Kozlowski, T. 1971. *Growth and Development of Trees*, Vols. 1 and 2. Academic Press, New York.

Kozlowski, T. T., P. J. Kramer, and S. G. Pallardy. 1991. *The Physiological Ecology of Woody Plants.* Academic Press, San Diego, CA.

Küppers, M. 1985. Carbon relations and competition between woody species in a Central European hedgerow. IV. Growth form and partitioning. *Oecologia* **66**: 343–52.

Küppers, M. 1989. Ecological significance of above-ground architectural patterns in woody plants: a question of cost-benefit relationships. *Trends Ecol. Evol.* **4**: 375–9.

Küppers, M. 1994. Canopy gaps: competitive light interception and economic space filling – a matter of whole-plant allocation. Pages 111–44 *in* M. M. Caldwell and R. W. Pearcy, eds. *Exploitation of Environmental Heterogeneity by Plants*: Ecophysiological Processes Above- and Belowground. Academic Press, San Diego, CA.

Kwesiga, F. R., J. F. Grace, and A. P. Sandford. 1986. Some photosynthetic characteristics of tropical timber trees as affected by light regime during growth. *Ann. Bot. (Lond.)* **58**: 23–32.

Larcher, W. 1983. *Physiological Plant Ecology*, 2nd ed. Springer-Verlag, Berlin.

Law, R. 1979. The cost of reproduction in annual meadow grass. *Am. Nat.* **113**: 3–16.

Lawton, J. H. 1995. Ecological Experiments with model systems. *Science* **269**: 328–31.

Lean, J. and D. A. Warrilow. 1989. Simulation of the regional climatic impact of Amazon deforestation. *Nature* **342**: 411–13.

Lechowicz, M. J. 1984. Why do temperate deciduous trees leaf out at different times? Adaptation and ecology of forest communities. *Am. Nat.* **124**: 821–42.

Lechowicz, M. J. and P. A. Blais. 1988. Assessing the contributions of multiple interacting traits to plant reproductive success: environmental dependence. *J. Evol. Biol.* **1**: 255–73.

Leck, M. A., V. T. Parker, and R. L. Simpson (eds.). 1989. *Ecology of Soil Seed Banks.* Academic Press, San Diego.

Lee, T. D. and F. A. Bazzaz. 1980. Effects of defoliation and competition on growth and reproduction in the annual plant *Abutilon theophrasti. J. Ecol.* **68**: 813–21.

Leemans, R. 1991. Sensitivity analysis of a forest succession model. *Ecol. Model.* 3: 247–62.

Levin, D. A. 1984. Inbreeding depression and proximity-dependent crossing success in *Phlox drummondii. Evolution* **38**: 116–27.

Levin, D. A. 1988. Plasticity, canalization and evolutionary stasis in plants. Pages 35–45 *in* A. J. Davy, M. J. Hutchings, and A. R. Watkinson, eds. *Plant Population Ecology.* Blackwell Scientific Publications, Oxford.

Levin, S. A. 1976. Population dynamic models in heterogeneous environments. *Annu. Rev. Ecol. Syst.* **7**: 287–310.

Levin, S. A. 1981. The role of theoretical ecology in the description and understanding of populations in heterogeneous environments. *Am. Zool.* **21**: 865–75.

Levin, S. A. 1993. Concepts of scale at the local level. Pages 7–19 *in* J. R. Ehleringer and C. B. Field, eds. *Scaling Physiological Processes. Leaf to Globe.* Academic Press, San Diego, CA.

Levins, R. 1968. *Evolution in Changing Environments*: Some Theoretical Explorations. Princeton University Press, Princeton, NJ.

Lewontin, R. C. 1957. The adaptation of populations to varying environments. *Cold Spring Harbor Symp. Quant. Biol.* **22**: 395–408.

Lieberman, M., D. Lieberman, and R. Peralta. 1989. Forests are not just swiss cheese: Canopy stereogeometry of non-gaps in tropical forests. *Ecology* **70**: 550–2.

Likens, G. E., F. H. Bormann, N. M. Johnson, D. W. Fisher, and R. S. Pierce. 1970. Effects of forest cutting and herbicide treatment on nutrient budgets in the Hubbard Brook Watershed-Ecosystem. *Ecol. Monogr.* **40**: 23–47.

Lincoln, D. E., E. D. Fajer, and R. H. Johnson. 1993. Plant-insect herbivore interactions in elevated CO_2 environments. *Trends Ecol. Evol.* **8**: 64–8.

Lindroth, R. L., K. K. Kinney, and C. L. Platz. 1993. Responses of deciduous trees to elevated atmospheric CO_2: productivity, phytochemistry, and insect performance. *Ecology* **74**: 763–77.

Lomnicki, A. 1988. *Population Ecology of Individuals.* Princeton University Press, Princeton, NJ.

Loucks, O. 1970. Evolution of diversity, efficiency and community stability. *Am. Zool.* **10**: 17–25.

Loveless, M. D. and J. L. Hamrick. 1984. Ecological determinants of genetic structure in plant populations. *Annu. Rev. Ecol. Syst.* **15**: 65–95.

Lovett, G. M. and J. D. Kinsman. 1990. Atmospheric deposition to high-elevation ecosystems. *Atmosph. Envir.* **24A**: 2767–86.

Lubchenco, J., A. M. Olson, L. B. Brubaker, S. R. Carpenter, M. M. Holland, S. P. Hubbell, S. A. Levin, J. A. MacMahon, P. A. Matson, J. M. Melillo, H. A. Mooney, C. H. Peterson, H. R. Pulliam, L. A. Real, P. J. Regal, and P. G. Risser. 1991. The sustainable biosphere initiative: An ecological research agenda. *Ecology* **72**: 371–412.

MacArthur, R. H. 1958. Population ecology of some warblers of northeastern coniferous forests. *Ecology* **39**: 599–619.

MacArthur, R. H. and E. O. Wilson. 1967. *The Theory of Island Biogeography*. Princeton University Press, Princeton, NJ.

Mack, R. N. 1986. Alien plant invasion into the intermountain West: a case history. Pages 191–213 *in* H. A. Mooney and J. A. Drake, eds. *Ecology of Biological Invasions of North America and Hawaii*. Springer-Verlag, New York.

MacMahon, J. A. 1981. Succesional processes: comparison among biomes with special reference to probable roles of and influences on animals. Pages 277–304 *in* D. C. West, H. H. Shugart, and D. B. Botkin, eds. *Forest Succession*: Concepts and Applications. Springer-Verlag, Berlin.

Mahall, B. E. and R. M. Callaway. 1991. Root communication among desert shrubs. *Proc. Natl. Acad. Sci. USA* **88**: 874–6.

Marks, P. L. and F. H. Bormann. 1972. Revegetation following forest cutting: mechanisms for return to steady-state nutrient cycling. *Science* **176**: 914–15.

Marks, P. L. and C. L. Mohler. 1985. Succession after elimination of buried seeds from a recently plowed field. *Bull. Torr. Bot. Club* **112**: 376–82.

Marschner, H. 1986. *Mineral Nutrition of Higher Plants*. Academic Press, London.

Marsden, J. E. and A. J. Tromba. 1981. *Vector Calculus*. W.H. Freeman, San Francisco, CA.

Martínez-Ramos, M., E. Alvarez-Buylla, J. Sarukhán, and D. Piñero. 1988*a*. Treefall age determination and gap dynamics in a tropical forest. *J. Ecol.* **76**: 700–16.

Martínez-Ramos, M., J. Sarukhán, and D. Piñero. 1988*b*. The demography of tropical trees in the context of forest gap dynamics. Pages 293–313 *in* A. J. Davy, M. J. Hutchings, and A. R. Watkinson, eds. *Plant Population Ecology*. Blackwell Scientific Publications, Oxford.

Matson, P. A. and P. M. Vitousek. 1981. Nitrification potentials following clearcutting in the Hoosier National Forest, Indiana. *Forest Sci.* **27**: 781–91.

Mattson, W. T. 1980. Herbivory in relation to plant nitrogen content. *Annu. Rev. Ecol. Syst.* **11**: 119–61.

May, R. and R. MacArthur. 1972. Niche overlap as a function of environmental variability. *Proc. Natl. Acad. Sci. USA* **69**: 1109–13.

May, R. M. 1973. On relationships among various types of populations models. *Am. Nat.* **107**: 46–57.

May, R. M. 1974. On the theory of niche overlap. *Theor. Pop. Biol.* **5**: 297–332.

May, R. M. (ed.) 1976. *Theoretical Ecology. Principles and Applications*. W. B. Saunders, Philadelphia, PA.

May, R. M. 1978. The evolution of ecological systems. *Sci. Am.* **239 (Sept.)**: 160–75.

McConnaughay, K. D. M. and F. A. Bazzaz. 1987. The relationship between gap size and performance of several colonizing annuals. *Ecology* **68**: 411–6.

McConnaughay, K. D. M. and F. A. Bazzaz. 1990. Interactions among colonizing annuals: Is there an effect of gap size? *Ecology* **71**: 1941–51.

McConnaughay, K. D. M. and F. A. Bazzaz. 1991. Is physical space a soil resource? *Ecology* **72**: 94–103.

McConnaughay, K. D. M. and F. A. Bazzaz. 1992*a*. The occupation and fragmentation of space: consequences of neighbouring roots. *Funct. Ecol.* **6**: 704–10.

McConnaughay, K. D. M. and F. A. Bazzaz. 1992*b*. The occupation and fragmentation of space: consequences of neighbouring shoots. *Funct. Ecol.* **6**: 711–18.

McDonnell, M. J. and E. W. Stiles. 1983. The structural complexity of old-field vegetation and the recruitment of bird-dispersed plant species. *Oecologia* **56**: 109–16.

McGee, A. B., M. R. Schmierbach, and F. A. Bazzaz. 1981. Photosynthesis and growth in populations of *Populus deltoides* from contrasting habitats. *Am. Midl. Nat.* **105**: 305–11.

McIntosh, R. P. 1980. The relationship between succession and the recovery process in ecosystems. Pages 11–62 *in* J. Cairns, ed. *The Recovery Process in Damaged Ecosystems*. Ann Arbor Scientific Publications, Ann Arbor, MI.

McIntosh, R. P. 1985. *The Background of Ecology*. Cambridge University Press, Cambridge.

McNeilly, T. and M. L. Roose. 1984. The distribution of perennial ryegrass genotypes in swards. *New Phytol.* **98**: 503–13.

Medina, E., L. Sternberg, and E. Cuevas. 1991. Vertical stratification of $\delta^{13}C$ values in closed natural and plantation forests in the Luquillo mountains, Puerto Rico. *Oecologia* **87**: 369–72.

Melillo, J. M. and J. R. Gosz. 1983. Interactions of the biogeochemical cycles in forest ecosystems. Pages 177–222 *in* B. Bolin and R. B. Cook, eds. *The Major Biogeochemical Cycles and Their Interactions*. Wiley & Sons, Chichester, UK.

Melillo, J. M., J. D. Aber, and J. F. Muratore. 1982. Nitrogen and lignin control of hardwood leaf litter decomposition dynamics. *Ecology* **63**: 621–6.

Melillo, J. M., A. D. McGuire, D. W. Kicklighter, B. Moore III, C. J. Vorosmarty, and A. L. Schloss. 1993. Global climate change and terrestrial net primary production. *Nature* **363**: 234–40.

Mellinger, M. V. and S. J. McNaughton. 1975. Structure and function of successional vascular plant communities in central New York. *Ecol. Monogr.* **45**: 161–82.

Miao, S. L. and F. A. Bazzaz. 1990. Responses to nutrient pulses of two colonizers requiring different disturbance frequencies. *Ecology* **71**: 2166–78.

Miao, S. L., F. A. Bazzaz, and R. B. Primack. 1991*a*. Effects of maternal nutrient pulse on reproduction of two colonizing *Plantago* species. *Ecology* **72**: 586–96.

Miao, S. L., F. A. Bazzaz, and R. B. Primack. 1991*b*. Persistence of maternal nutrient effects in *Plantago major*: the third generation. *Ecology* **72**: 1634–42.

Michaels, H. J. and F. A. Bazzaz. 1986. Resource allocation and demography of sexual and apomictic *Antennaria parlinii*. *Ecology* **67**: 27–36.

Michaels, H. J. and F. A. Bazzaz. 1989. Individual and population responses of sexual and apomictic plants to environmental gradients. *Am. Nat.* **134**: 190–207.

Miles, J. 1979. *Vegetation Dynamics*. Chapman & Hall, London.

Miles, J. 1987. Vegetation succession: past and present perceptions. Pages 1–29 *in* A. J. Gray, M. J. Crawley, and P. J. Edwards, eds. *Colonization, Succession and Stability*. Blackwell Scientific Publications, Oxford.

Miller, T. E. and P. A. Werner. 1987. Competitive effects and responses between plant species in a first-year old-field community. *Ecology* **68**: 1201–10.

Mitchell-Olds, T. 1992. Does environmental variation maintain genetic variation? A question of scale. *Trends Ecol. Evol.* **7**: 397–8.

Mooney, H. A. 1972. The carbon balance of plants. *Annu. Rev. Ecol. Syst.* **3**: 315–46.

Mooney, H. A. 1976. Some contributions of physiological ecology to plant population biology. *Syst. Bot.* **1**: 269–83.

Mooney, H. A. 1991. Plant physiological ecology – determinants of progress. *Funct. Ecol.* **5**: 127–35.

Mooney, H. A. and N. Chiariello. 1984. The study of plant function: the plant as a balanced system. Pages 305–23 *in* R. Dirzo and J. Sarukhán, eds. *Perspectives in Plant Population Biology.* Sinauer Associates, Sunderland, MA.

Mooney, H. A. and J. R. Drake (eds.). 1986. *Ecology of Biological Invasions of North America and Hawaii.* Springer-Verlag, Berlin.

Mooney, H. A. and M. Godron (eds.). 1983. *Disturbance and Ecosystems.* Springer-Verlag, Berlin.

Mooney, H. A. and S. L. Gulmon. 1982. Constraints on leaf structure and function in reference to herbivory. *BioScience* **32**: 198–206.

Mooney, H. A., C. Field, S. L. Gulmon, and F. A. Bazzaz. 1981. Photosynthetic capacity in relation to leaf position in desert versus old-field annuals. *Oecologia* **50**: 109–12.

Mooney, H., R. Pearcy, and J. Ehleringer. 1987a. Plant physiological ecology today. *BioScience* **37**: 18–20.

Mooney, H. A., P. M. Vitousek, and P. A. Matson. 1987b. Exchange of materials between terrestrial ecosystems and the atmosphere. *Science* **238**: 926–32.

Mooney, H. A., B. G. Drake, R. J. Luxmoore, W. C. Oechel, and L. F. Pitelka. 1991. Predicting ecosystem responses to elevated CO_2 concentrations. *BioScience* **41**: 96–104.

Mousseau, M. and B. Saugier. 1992. The direct effect of increased carbon on gas dioxide exchange and growth of forest tree species. *J. Exp. Bot.* **43**: 1121–30.

Mueller-Dombois, D. 1992. Potential effects of the increase in carbon dioxide and climate change on the dynamics of vegetation. Pages 61–79 *in* J. Wisniewski and A. E. Lugo, eds. *Natural Sinks of CO_2.* Kluwer Academic Publishers, Dordrecht.

Muller, R. N. and F. H. Bormann. 1976. Role of *Erythronium americanum* Ker. in energy flow and nutrient dynamics of a northern hardwood forest ecosystem. *Science* **193**: 1126–8.

Newell, E. A. 1991. Direct and delayed costs of reproduction in *Aesculus californica. J. Ecol.* **79**: 365–78.

Newell, E. A., E. P. McDonald, B. R. Strain, and J. S. Denslow. 1993. Photosynthetic responses of *Miconia* species to canopy openings in a lowland tropical rainforest. *Oecologia* **94**: 49–56.

Newman, E. I. 1982. Niche separation and species diversity in terrestrial vegetation. Pages 61–78 *in* E. I. Newman, ed. *The Plant Community as a Working Mechanism.* Blackwell Scientific Publications, Oxford.

Niklas, K. 1988. The role of phyllotactic pattern as a 'developmental constraint' on the interception of light by leaf surfaces. *Evolution* **42**: 1–16.

Nobel, P. S. 1991. *Physicochemical and Environmental Plant Physiology.* Academic Press, San Diego, CA.

Noble, I. R. and R. O. Slatyer. 1980. The use of vital attributes to predict successional changes in plant communities subject to recurrent disturbances. *Vegetatio* **43**: 5–21.

Norman, J. M. 1993. Scaling processes between leaf and canopy levels. Pages 41–76 *in* J. R. Ehleringer and C. B. Field, eds. *Scaling Physiological Processes. Leaf to Globe.* Academic Press, San Diego, CA.

Novoplansky, A., D. Cohen, and T. Sachs. 1990. How *Portulaca* seedlings avoid their neighbours. *Oecologia* **82**: 490–3.

Numata, M. 1990. *Ecology and Conservation.* Meiseikai, Tokyo.

Numata, M. and H. Yamai. 1955. The developmental process of weed communities – experimental studies on early stages of secondary succession, I. *Jap. J. Ecol.* **4**: 166–71.

O'Neill, R. V., D. L. DeAngelis, J. B. Waide, and T. F. H. Allen. 1986. *A Hierarchical Concept of Ecosystems*. Princeton University Press, Princeton, NJ.

Oberbauer, S., and B. Strain. 1985. Effects of light regime on the growth and physiology of *Pentaclethra macroloba* (Mimosaceae) in Costa Rica. *J. Trop. Ecol.* **1**: 303–20.

Odum, E. P. 1960. Organic production and turnover in old field succession. *Ecology* **41**: 34–49.

Odum, E. P. 1969. The strategy of ecosystem development. *Science* **164**: 262–70.

Odum, E. P. 1993. *Ecology and Our Endangered Life Support Systems*. Sinauer Associates, Sunderland, MA.

Oechel, W. C., S. Cowles, N. Grulke, S. J. Hastings, B. Lawrence, T. Prudhomme, G. Riechers, B. Strain, D. Tissue, and G. Vourlitis. 1994. Transient nature of CO_2 fertilization in Arctic tundra. *Nature* **371**: 500–3.

Oldeman, R. A. A. 1978. Architecture and energy exchange of dicotyledonous trees in the forest. Pages 535–60 *in* P. B. Tomlinson and M. H. Zimmerman, eds. *Tropical Trees as Living Systems*. Cambridge University Press, Cambridge.

Olson, J. S. 1958. Rates of succession and soil changes on southern Lake Michigan sand dunes. *Bot. Gazette* **119**: 125–70.

Oosting, H. J. 1942. An ecological analysis of the plant communities of Piedmont, North Carolina. *Am. Midl. Nat.* **28**: 1–126.

Oosting, H. J. 1956. *The Study of Plant Communities*. W. H. Freeman, San Francisco, CA.

Orians, G. H. 1983. The influence of tree-falls in tropical forests on tree species richness. *Trop. Ecol.* **23**: 255–79.

Ormsbee, P., F. A. Bazzaz, and W. R. Boggess. 1976. Physiological ecology of *Juniperus virginiana* in oldfields. *Oecologia* **23**: 75–82.

Osbornová, J., M. Kovárová, J. Leps, and K. Prach (eds.). 1990. *Succession in Abandoned fields. Studies in Central Bohemia, Czechoslovakia*. Kluwer Academic Publishers, Dordrecht.

Osmond, C. B., O. Björkman, and D. J. Anderson. 1980. *Physiological Processes in Plant Ecology*: Toward a Synthesis with Atriplex. Springer-Verlag, Berlin.

Osunkoya, O. O. and J. E. Ash. 1991. Acclimation to a change in light regime in seedlings of six Australian rainforest tree species. *Aust. J. Bot.* **39**: 591–605.

Owensby, C. E. 1993. Potential impacts of elevated CO_2 and above and below ground litter quality of a tall grass prairie. *Water, Air and Soil Pollution* **70**: 413–24.

Pacala, S. W., C. D. Canham, and J. A. Silander, Jr. 1993. Forest models defined by field measurements: I. The design of a northeastern forest simulator. *Can. J. Forest Sci.* **23**: 1980–8.

Parker, V. T., R. L. Simpson, and M. A. Leck. 1989. Pattern and process in the dynamics of seed banks. Pages 367–84 *in* M. A. Leck, V. T. Parker, and R. L. Simpson, eds. *Ecology of Soil Seed Banks*. Academic Press, San Diego, CA.

Parrish, J. A. D. and F. A. Bazzaz. 1976. Underground niche separation in successional plants. *Ecology* **57**: 1281–8.

Parrish, J. A. D. and F. Bazzaz. 1978. Pollination niche separation in a winter annual community. *Oecologia* **35**: 133–40.

Parrish, J. A. D. and F. A. Bazzaz. 1979. Difference in pollination niche relationships in early and late successional plant communities. *Ecology* **60**: 597–610.

Parrish, J. A. D. and F. A. Bazzaz. 1982*a*. Competitive interactions in plant communities of different successional ages. *Ecology* **63**: 314–20.

Parrish, J. A. D. and F. A. Bazzaz. 1982*b*. Niche responses of early and late successional tree seedlings on three resource gradients. *Bull. Torr. Bot. Club* **109**: 451–6.

Parrish, J. A. D. and F. A. Bazzaz. 1982*c*. Responses of plants from three successional communities to a nutrient gradient. *J. Ecol.* **70**: 233–48.

Parrish, J. A. D. and F. A. Bazzaz. 1985*a*. Ontogenetic niche shifts in old-field annuals. *Ecology* **66**: 1296–302.

Parrish, J. A. D. and F. A. Bazzaz. 1985*b*. Nutrient content of *Abutilon theophrasti* seeds and the competitive ability of the resulting plants. *Oecologia* **65**: 247–51.

Pastor, J. and W. M. Post. 1988. Response of northern forests to CO_2–induced climatic change. *Nature* **334**: 55–8.

Pastor, J., M. A. Stillwell, and D. Tilman. 1987. Nitrogen mineralization and nitrification in four Minnesota old fields. *Oecologia* **71**: 481–5.

Paulson, T. L. and W. J. Platt. 1989. Gap light regimes influence canopy tree diversity. *Ecology* **70**: 553–5.

Pearcy, R. W. 1987. Photosynthetic gas exchange responses of Australian tropical forest trees in canopy, gap and understory. *Funct. Ecol.* **1**: 169–78.

Pearcy, R. W. 1988. Photosynthetic utilisation of lightflecks by understory plants. *Aust. J. Plant Physiol.* **15**: 223–38.

Pearcy, R. W. 1990. Sunflecks and photosynthesis in plant canopies. *Annu. Rev. Plant Physiol.* **41**: 421–53.

Pearcy, R. W. and D. A. Sims. 1994. Photosynthetic acclimation to changing light environments: scaling from the leaf to the whole plant. Pages 145–74 *in* M. M. Caldwell and R. W. Pearcy, eds. *Exploitation of Environmental Heterogeneity by Plants*: Ecophysiological Processes Above- and Belowground. Academic Press, San Diego, CA.

Pearcy, R. W., J. Ehleringer, H. A. Mooney, and P. W. Rundel. 1989. *Plant Physiological Ecology. Field Methods and Instrumentation*. Chapman & Hall, London.

Pearcy, R. W., R. L. Chazdon, L. J. Gross, and K. A. Mott. 1994. Photosynthetic utilization of sunflecks: a temporally patchy resource on a time scale of seconds to minutes. Pages 175–208 *in* M. M. Caldwell and R. W. Pearcy, eds. *Exploitation of Environmental Heterogeneity by Plants*: Ecophysiological Processes Above- and Belowground. Academic Press, San Diego, CA.

Peet, R. K. and N. L. Christensen. 1980. Succession: a population process. *Vegetatio* **43**: 131–40.

Perozzi, R. E. and F. A. Bazzaz. 1978. The response of an early successional community to shortened growing season. *Oikos* **31**: 89–93.

Peterjohn, W. T., J. M. Melillo, F. P. Bowles, and P. A. Steudler. 1993. Soil warming and trace gas fluxes: experimental design and preliminary flux results. *Oecologia* **93**: 18–24.

Peterson, C. J. and S. T. A. Pickett. 1990. Microsite and elevational influences on early forest regeneration after catastrophic windthrow. *J. Veget. Sci.* **1**: 657–62.

Peterson, D. L. and F. A. Bazzaz. 1978. Life cycle characteristics of *Aster pilosus* in early successional habitats. *Ecology* **59**: 1005–13.

Petraitis, P. S. 1981. Algebraic and graphical relationships among niche breadth measures. *Ecology* **62**: 545–8.

Pianka, E. R. 1994. *Evolutionary Ecology*, 5th ed. Harper-Collins, New York.
Pickett, S. T. A. 1976. Succession: An evolutionary interpretation. *Am. Nat.* **110**: 107–19.
Pickett, S. T. A. 1980. Non-equilibrium coexistence of plants. *Bull. Torrey Bot. Club* **107**: 238–48.
Pickett, S. T. A. 1989. Space-for-time substitution as an alternative to long-term studies. Pages 110–35 *in* G. E. Likens, ed. *Long-term Studies in Ecology. Approaches and Alternatives.* Springer-Verlag, New York.
Pickett, S. T. A. and F. A. Bazzaz. 1976. Divergence of two co-occurring successional annuals on a soil moisture gradient. *Ecology* **57**: 169–76.
Pickett, S. T. A. and F. A. Bazzaz. 1978*a*. Germination of co-occurring annual species on a soil moisture gradient. *Bull. Torrey Bot. Club* **105**: 312–6.
Pickett, S. T. A. and F. A. Bazzaz. 1978*b*. Organization of an assemblage of early successional species on a soil moisture gradient. *Ecology* **59**: 1248–55.
Pickett, S. T. A. and P. S. White (eds.). 1985. *The Ecology of Natural Disturbance and Patch Dynamics.* Academic Press, London.
Pickett, S. T. A., S. L. Collins, and J. J. Armesto. 1987. Models, mechanisms and pathways of succession. *Bot. Rev.* **53**: 335–71.
Pickett, S. T. A., J. Kolasa, J. J. Armesto, and S. L. Collins. 1989. The ecological concept of disturbance and its expression at various hierarchical levels. *Oikos* **54**: 129–36.
Pimm, S. L., G. E. Davis, L. Loope, C. T. Roman, T. J. I. Smith, and J. T. Tilmant. 1994. Hurricane Andrew. *BioScience* **44**: 224–9.
Piñero, D., M. Martínez-Ramos, A. Mendoza, E. Alvarez-Buylla, and J. Sarukhán. 1986. Demographic studies in *Astrocaryum mexicanum* and their use in understanding community dynamics. *Principes* **30**: 108–16.
Pitelka, L. F. and D. J. Raynal. 1989. Forest decline and acidic deposition. *Ecology* **70**: 2–10.
Popma, J. and F. Bongers. 1988. The effect of canopy gaps on growth and morphology of seedlings of rain forest species. *Oecologia* **75**: 625–32.
Popma, J. and F. Bongers. 1991. Acclimation of seedlings of three tropical rain forest species to changing light availability. *J. Trop. Ecol.* **7**: 85–97.
Prentice, I. C. 1993. Climate change – process and production. *Nature* **363**: 209–10.
Prentice, I. C., P. J. Bartlein, and T. Webb III. 1991. Vegetation and climate change in eastern North America since the last glacial maximum. *Ecology* **72**: 2038–56.
Prentice, I. C., M. T. Sykes, and W. Cramer. 1993. A simulation model for the transient effects of climate change on forest landscapes. *Ecol. Model.* **65**: 51–70.
Primack, R. B. and P. Hall. 1990. Costs of reproduction in the pink lady's slipper orchid: a four-year experimental study. *Am. Nat.* **136**: 638–56.
Quarterman, E. 1957. Early plant succession on abandoned cropland in the Central Basin of Tennessee. *Ecology* **38**: 300–9.
Rastetter, E. B. and G. R. Shaver. 1992. A model of multiple-element limitation for acclimating vegetation. *Ecology* **73**: 1157–74.
Rastetter, E. B., A. W. King, B. J. Cosby, G. M. Hornberger, R. V. O'Neill, and J. E. Hobbie. 1992. Aggregating fine-scale ecological knowledge to model coarser-scale attributes of ecosystems. *Ecol. Applic.* **2**: 55–70.
Rathcke, B. and E. P. Lacey. 1985. Phenological patterns of terrestrial plants. *Annu. Rev. Ecol. Syst.* **16**: 179–214.

Raynal, D. J. and F. A. Bazzaz. 1973. Establishment of early successional plant populations on forest and prairie soils. *Ecology* **54**: 1335–41.

Raynal, D. J. and F. A. Bazzaz. 1975a. The contrasting life-cycle strategies of three summer annuals found in abandoned fields in Illinois. *J. Ecol.* **63**: 587–96.

Raynal, D. J. and F. A. Bazzaz. 1975b. Interference of winter annuals with *Ambrosia artemisiifolia* in early successional fields. *Ecology* **56**: 35–49.

Reekie, E. G. and F. A. Bazzaz. 1987a. Reproductive effort in plants. 1. Carbon allocation to reproduction. *Am. Nat.* **129**: 876–96.

Reekie, E. G. and F. A. Bazzaz. 1987b. Reproductive effort in plants. 2. Does carbon reflect the allocation of other resources? *Am. Nat.* **129**: 897–906.

Reekie, E. G. and F. A. Bazzaz. 1987c. Reproductive effort in plants. 3. Effect of reproduction on vegetative activity. *Am. Nat.* **129**: 907–19.

Reekie, E. G. and F. A. Bazzaz. 1992. Cost of reproduction as reduced growth in genotypes of two congeneric species with constrasting life histories. *Oecologia* **90**: 21–6.

Regehr, D. L. and F. A. Bazzaz. 1976. Low temperature photosynthesis in successional winter annuals. *Ecology* **57**: 1297–303.

Regehr, D. L. and F. A. Bazzaz. 1979. The population dynamics of *Erigeron canadensis*, a successional winter annual. *J. Ecol.* **67**: 923–33.

Reich, P. B., M. B. Walters, and D. S. Ellsworth. 1992. Leaf life-span in relation to leaf, plant, and stand characteristics among diverse ecosystems. *Ecol. Monogr.* **62**: 365–92.

Reiners, W. A. 1992. Twenty years of ecosystem reorganization following experimental deforestation and regrowth suppression. *Ecol. Monogr.* **62**: 503–23.

Reynolds, J. F., B. Acock, R. L. Dourgherty, and J. D. Tenhunen. 1989. A modular structure for plant growth simulation models. Pages 000–00 *in* J. S. Pereira and J. J. Landsberg, eds. *Biomass Production by Fast-Growing Trees.* Kluwer Academic Publishers, Boston, MA.

Reynolds, J. F., D. W. Hilbert, and P. R. Kemp. 1993. Scaling ecophysiology from the plant to the ecosystem: a conceptual framework. Pages 127–40 *in* J. R. Ehleringer and C. B. Field, eds. *Scaling Physiological Processes. Leaf to Globe.* Academic Press, San Diego, CA.

Rice, E. L. 1984. *Allelopathy*, 2nd ed. Academic Press, Orlando, FL.

Rice, E. L. and S. K. Pancholy. 1972. Inhibition of nitrification by climax ecosystems. *Am J Bot* **59**: 1033–40.

Rice, K. and S. Jain. 1985. Plant population genetics and evolution in disturbed environments. Pages 287–303 *in* S. T. A. Pickett and P. S. White, eds. *The Ecology of Natural Disturbance and Patch Dynamics*. Academic Press, London.

Rice, S. and F. A. Bazzaz. 1989a. Quantification of plasticity of plant traits in response to light intensity: Comparing phenotypes at a common weight. *Oecologia* **78**: 502–7.

Rice, S. A. and F. A. Bazzaz. 1989b. Growth consequences of plasticity of plant traits in response to light conditions. *Oecologia* **78**: 508–12.

Richards, P. W. 1952. *The Tropical Rain Forest*: An Ecological Study. Cambridge University Press, London.

Richards, P. W. 1995. *The Tropical Rain Forest*: An Ecological Study, 2nd ed. Cambridge University Press, Cambridge.

Ricklefs, R. 1977. Environmental heterogeneity and plant species diversity: a hypothesis. *Am. Nat.* **111**: 376–81.

Ricklefs, R. 1990. *Ecology*, 3rd ed. W. H. Freeman, New York.

Roach, D. A. and R. D. Wulff. 1987. Maternal effects in plants. *Annu. Rev. Ecol. Syst.* **18**: 20935.

Roberts, E. H. 1981. The interaction of environmental factors controlling loss of dormancy in seeds. *Ann. Appl. Biol.* **98**: 552–5.

Robertson, G. P. 1984. Nitrification and nitrogen mineralization in a lowland rainforest succession in Costa Rica, Central America. *Oecologia* **61**: 99–104.

Robertson, G. P. and P. M. Vitousek. 1981. Nitrification potentials in primary and secondary succession. *Ecology* **62**: 376–86.

Roos, F. H. and J. A. Quinn. 1977. Phenology and reproductive allocation in *Andropogon scoparius* (Graminae) populations in communities of different successional stages. *Am. J. Bot.* **64**: 535–40.

Root, R. B. 1967. The niche exploitation pattern of the blue-grey gnatcatcher. *Ecol. Monogr.* **37**: 317–50.

Rosenzweig, M. L. 1991. Habitat selection and population interactions: the search for mechanism. *Am. Nat.* **137**: S5–S28.

Roughgarden, J. 1979. *Theory of Population Genetics and Evolutionary Ecology*: An Introduction. Macmillan, New York.

Runkle, J. R. 1985. Disturbance regimes in temperate forests. Pages 17–34 *in* S. T. A. Pickett and P. S. White, eds. *The Ecology of Natural Disturbance and Patch Dynamics*. Academic Press, London.

Running, S. W. and E. R. Hunt, Jr. 1993. Generalization of a forest ecosystem process model for other biomes, BIOME-BGC, and an application for global-scale models. Pages 141–58 *in* J. R. Ehleringer and C. B. Field, eds. *Scaling Physiological Processes. Leaf to Globe*. Academic Press, San Diego, CA.

Russell, G., B. Marshall, and P. G. Jarvis. 1989. *Plant Canopies*: Their Growth, Form and Function. Cambridge University Press, Cambridge.

Sagan, C. and R. Turco. 1990. *A Path Where No Man Thought. Nuclear Winter and the End of the Arms Race*. Random House, New York.

Salati, E., J. Marques, and L. C. B. Molion. 1978. Origem e distribuição das chuvas na Amazônia. *Interciencia* **3(4)**: 200–6.

Salisbury, F. B. and C. W. Ross. 1985. *Plant Physiology*. Wadsworth, Belmont, CA.

Salzman, A. G. 1985. Habitat selection in a clonal plant. *Science* **228**: 603–4.

Salzman, G. and M. A. Parker. 1985. Neighbors ameliorate local salinity stress for a rhizomatous plant in a heterogeneous environment. *Oecologia* **65**: 273–7.

Sarmiento, J. L. and E. T. Sundquist. 1992. Revised budget for the oceanic uptake of anthropogenic carbon dioxide. *Nature* **356**: 589–93.

Sarukhán, J., M. Martínez-Ramos, and D. Piñero. 1984. The analysis of demographic variability at the individual level and its populational consequences. Pages 83–106 *in* R. Dirzo and J. Sarukhán, eds. *Perspectives in Plant Population Biology*. Sinauer Associates, Sunderland, MA.

Sasek, T. W. and B. R. Strain. 1990. Implications of atmospheric carbon dioxide enrichment and climatic change for the geographical distribution of two introduced vines in the USA. *Climate Change* **16**: 31–52.

Schaal, B. A. 1984. Life-history variation, natural selection, and maternal effects in plant populations. Pages 188–206 *in* R. Dirzo and J. Sarukhán, eds. *Perspectives in Plant Population Ecology*. Sinauer Associates, Sunderland, MA.

Schaal, B. A. 1985. Genetic variation in plant populations: From demography to DNA. Pages 321–41 *in* J. Haeck and J. W. Waldendorp, eds. *Structure and*

Functioning of Plant Populations. 2. Phenotypic and Genotypic Variation in Plant Populations. North-Holland Publishing Company, Amsterdam.

Schaetzl, R. J., S. F. Burns, D. L. Johnson, and T. W. Small. 1989. Tree uprooting: review of impacts on forest ecology. *Vegetatio* **79**: 165–76.

Schimel, J. P., L. E. Jackson, and M. K. Firestone. 1989. Spatial and temporal effects on plant-microbial competition for inorganic nitrogen in a California annual grassland. *Soil Biol. Biochem.* **21**: 1059–66.

Schlichting, C. D. 1986. The evolution of phenotypic plasticity in plants. *Annu. Rev. Ecol. Syst.* **17**: 667–93.

Schmalhausen, I. I. 1949. *Factors of Evolution*: The Theory of Stabilizing Selection. University of Chicago Press, Chicago.

Schmid, B. 1992. Phenotypic variation in plants. *Evol. Trends in Plants* **6**: 45–60.

Schmid, B. and F. A. Bazzaz. 1987. Clonal integration and population structure in perennials: Effects of severing rhizome connections. *Ecology* **68**: 2016–22.

Schmid, B. and F. A. Bazzaz. 1990. Plasticity in plant size and architecture in rhizome-derived vs. seed-derived *Solidago* and *Aster*. *Ecology* **71**: 523–35.

Schmid, B. and F. A. Bazzaz. 1992. Growth responses of rhizomatous plants to fertilizer application and interference. *Oikos* **65**: 13–24.

Schmid, B. and F. A. Bazzaz. 1994. Crown construction, leaf dynamics, and carbon gain in two perennials with contrasting architecture. *Ecol. Monogr.* **64**: 177–203.

Schmid, B. and J. L. Harper. 1985. Clonal growth in grassland perennials. I. Density and pattern-dependent competition between plants with different growth forms. *J. Ecol.* **73**: 793–808.

Schmid, B., and J. Stöklin (eds.). 1991. *Populationsbiologie der Pflanzen*. Birkhäuser, Basel.

Schmid, B., G. M. Puttick, K. Burgess, and F. A. Bazzaz. 1988. Clonal integration and effects of simulated herbivory in old-field perennials. *Oecologia* **75**: 465–71.

Schmid, B., S. L. Miao, and F. A. Bazzaz. 1990. Effects of simulated root herbivory and fertilizer application on growth and biomass allocation in the clonal perennial *Solidago canadensis*. *Oecologia* **84**: 9–15.

Schmitt, J., J. Niles, and R. D. Wulff. 1992. Norms of reaction of seed traits to maternal environments in *Plantago lanceolata*. *Am. Nat.* **139**: 451–66.

Schulze, E.-D. 1982. Plant life forms and their carbon, water, and nutrient relations. Pages 615–76 *in* O. L. Lange, P. S. Nobel, C. B. Osmond, and H. Ziegler, eds. *Plant Physiological Ecology, Vol. II. Water Relations and Carbon Assimilation*. Springer-Verlag, Berlin.

Schulze, E.-D. 1986. Carbon dioxide and water vapor exchange in response to drought in the atmosphere and the soil. *Annu. Rev. Plant Physiol.* **37**: 247–74.

Schulze, E.-D. 1989. Air pollution and forest decline in a spruce (*Picea abies*) forest. *Science* **244**: 776–83.

Schulze, E.-D., F. M. Kelliher, C. Körner, J. Lloyd, and R. Leuning. 1994. Relationships among maximum stomatal conductance, ecosystem surface conductance, carbon assimilation rate, and plant nitrogen nutrition: a global ecology scaling exercise. *Annu. Rev. Ecol. Syst.* **25**: 629–60.

Schwaegerle, K. E. and F. A. Bazzaz. 1987. Differentiation among nine populations of *Phlox*. II. Response to environmental gradients. *Ecology* **68**: 54–64.

Schwartz, D. M. and F. A. Bazzaz. 1973. *In situ* measurements of carbon dioxide gradients in a soil–plant–atmosphere system. *Oecologia* **12**: 161–7.

Sellers, P. J. 1987. Modelling effects of vegetation on climate. Pages 133–62 *in* R. E. Dickinson, ed. *The Geophysiology of Amazonia.* Wiley & Sons, New York.

Shaver, G. R., W. D. Billings, F. S. Chapin, A. E. Giblin, K. J. Nadelhoffer, W. C. Oechel, and E. B. Rastetter. 1992. Global change and the carbon balance of arctic ecosystems. *BioScience* **42**: 433–41.

Shipley, B. and R. H. Peters. 1990. A test of the Tilman model of plant strategies: relative growth rate and biomass partitioning. *Am. Nat.* **136**: 139–53.

Shmida, A. and S. P. Ellner. 1984. Coexistence of plants with similar niches. *Vegetatio* **58**: 29–55.

Shugart, H. H. 1984. *A Theory of Forest Dynamics*: The Ecological Implications of Forest Succession Models. Springer-Verlag, Berlin.

Shugart, H. H. 1990. Using ecosystem models to assess potential consequences of global climate change. *Trends Ecol. Evol.* **5**: 303–7.

Shugart, H. H. and W. R. Emanuel. 1985. Carbon dioxide increase: The implications at the ecosystem level. *Plant Cell Env.* **8**: 381–6.

Shugart, H. H., D. C. West, and W. R. Emmanuel. 1981. Patterns and dynamics of forest: an application of simulation models. Pages 77–94 *in* D. C. West, H. H. Shugart, and D. B. Botkin, eds. *Forest Succession*: Concepts and Applications. Springer-Verlag, Berlin.

Shukla, J., C. Nobre and P. Sellers. 1990. Amazon deforestation and climate change. *Science* **247**: 1322–5.

Silvertown, J. W. 1987. *Introduction to Plant Population Biology*. Longman, London.

Silvertown, J. W. and R. Law. 1987. Do plants need niches? Some recent developments in plant community ecology. *Trends Ecol. Evol.* **2**: 24–6.

Silvertown, J. W. and J. Lovett Doust. 1993. *Introduction to Plant Population Biology*. Blackwell Scientific Publications, Oxford.

Sims, D. A. and R. W. Pearcy. 1991. Photosynthesis and respiration in *Alocassia macrorrhiza* following transfers to high and low light. *Oecologia* **86**: 447–53.

Sipe, T. W. 1990. Gap Partitioning Among Maples (*Acer*) in the Forests of Central New England. Ph. D. Thesis. Harvard University, Cambridge, MA.

Sipe, T. W. and F. A. Bazzaz. 1994. Gap partitioning among maples (*Acer*) in Central New Engand: shoot architecture and photosynthesis. *Ecology* **75**: 2318–32.

Sipe, T. W. and F. A. Bazzaz. 1995. Gap partitioning among maples (*Acer*) in Central New Engand: survival and growth. *Ecology* **76**: 1587–602.

Smith, E. P. 1982. Niche breadth, resource availability and inference. *Ecology* **63**: 1675–81.

Smith, T. M. and M. A. Huston. 1989. A theory of the spatial and temporal dynamics of plant communities. *Vegetatio* **83**: 49–69.

Smith, T. M. and H. H. Shugart. 1993. The transient response of terrestrial carbon storage to a perturbed climate. *Nature* **361**: 523–6.

Smith, T. M. and D. L. Urban. 1988. Scale and the resolution of forest structural pattern. *Vegetatio* **74**: 143–50.

Smith, T. M., W. P. Cramer, R. K. Dixon, R. Leemans, R. P. Nielson, and A. M. Solomon. 1993. The global terrestrial carbon cycle. *Water Air and Soil Pollution* **70**: 19–37.

Smith, T. M., F. I. Woodward, and H. H. Shugart (eds.). 1995. *Plant Functional Types*. Cambridge University Press, Cambridge.

Smith, W. K., A. K. Knapp, and W. A. Reiners. 1989. Penumbral effects on sunlight penetration in plant communities. *Ecology* **70**: 1603–9.

Snaydon, R. W. and M. S. Davies. 1972. Rapid population differentiation in a mosaic environment. II. Morphological variation in *Anthoxanthum odoratum*. *Evolution* **26**: 390–405.

Snow, A. A. and D. F. Whigham. 1989. Costs of flower and fruit production in *Tipularia discolor* (Orchidaceae). *Ecology* **70**: 1286–93.

Solbrig, O. T. 1971. The population biology of dandelions. *Am. Sci.* **59**: 686–94.

Solbrig, O. T. 1981. Studies on the population biology of the genus *Viola*. II. The effect of plant size on fitness in *Viola sororia*. *Evolution* **35**: 1080–93.

Solomon, A. M. 1986. Transient response of forests to CO_2–induced climate change: simulation modeling experiments in eastern North America. *Oecologia* **68**: 567–79.

Sorrenson-Cothern, K. A., E. D. Ford, and D. G. Sprugel. 1993. A model of competition incorporating plasticity through modular foliage and crown development. *Ecol. Monogr.* **63**: 277–304.

Sousa, W. P. 1979. Experimental investigations of disturbance and ecological succession in a rocky intertidal algal community. *Ecol. Monogr.* **49**: 227–54.

Sousa, W. P. 1984. The role of disturbance in natural communities. *Annu. Rev. Ecol. Syst.* **15**: 353–91.

Spieth, P. T. 1979. Environmental heterogeneity: a problem of contradictory selection pressures, gene flow, and local polymorphism. *Am. Nat.* **113**: 247–60.

Stanton, M. L. 1984*a*. Seed variation in wild radish: effect of seed size on components of seedling and adult fitness. *Ecology* **65**: 1105–12.

Stanton, M. L. 1984*b*. Developmental and genetic sources of seed weight variation in *Raphanus raphanistrum* L. (Brassiceae). *Am. J. Bot.* **71**: 1090–8.

Stephens, G. R. and P. E. Waggoner. 1970. The forests anticipated from 40 years of natural transitions in mixed hardwoods. *Bull. Conn. Agric. Exper. Station New Haven*, No. 707.

Strain, B. R. 1987. Direct effects of increasing atmospheric CO_2 on plants and ecosystems. *Trends Ecol. Evol.* **2**: 18–19.

Strain, B. R. 1991. Possible genetic effects of continually increasing atmospheric CO_2. Pages 237–44 *in* G. E. Taylor, Jr, L. F. Pitelka, and M. T. Clegg, eds. *Ecological Genetics and Air Pollution*. Springer-Verlag, New York.

Strain, B. R. and J. D. Cure (eds.). 1985. *Direct Effects of Increasing Carbon Dioxide on Vegetation*. United States Department of Energy, Carbon Dioxide Research Division, Office of Energy Research, Washington, DC.

Strauss-Debenedetti, S. and F. A. Bazzaz. 1991. Plasticity and acclimation to light in tropical Moraceae of different successional positions. *Oecologia* **87**: 377–87.

Strauss-Debenedetti, S. and F. A. Bazzaz. 1996, Photosynthetic characteristics of tropical trees along successional gradients. Pages 162–86 *in* S. S. Mulkey, R. Chazdon, A. P. Smith (eds). *Tropical Forest Plant Ecophysiology*. Chapman & Hall, New York.

Strong, D. R. Jr. 1983. Natural variability and the manifold mechanisms of ecological communities. *Am. Nat.* **122**: 636–60.

Sultan, S. E. 1987. Evolutionary implications of phenotypic plasticity in plants. *Evol. Biol.* **21**: 127–78.

Sultan, S. E. 1993. Phenotypic plasticity and the Neo-Darwinian legacy. *Evol. Trends Plants* **6**: 61–71.

Sultan, S. E. and F. A. Bazzaz. 1993*a*. Phenotypic plasticity in *Polygonum persicaria*. I. Diversity and uniformity in genotypic norms of reaction to light. *Evolution* **47**: 1009–31.

Sultan, S. E. and F. A. Bazzaz. 1993*b*. Phenotypic plasticity in *Polygonum persicaria*. II. Norms of reaction to soil moisture and the maintenance of genetic diversity. *Evolution* **47**: 1032–49.

Sultan, S. E. and F. A. Bazzaz. 1993*c*. Phenotypic plasticity in *Polygonum persicaria*. III. The evolution of ecological breadth for nutrient environment. *Evolution* **47**: 1050–71.

Tans, P. P., I. Y. Fung, and T. Takahashi. 1990. Observational constraints on the global atmospheric CO_2 budget. *Science* **247**: 1431–8.

Tans, P. P., I. Y. Fung, and I. G. Enting. 1995. Storage versus flux budgets: the terrestrial uptake of CO_2 during the 1980s. Pages 351–66 *in* G. M. Woodwell and F. T. Mackenzie, eds. *Biotic Feedbacks in the Global Climatic System. Will the Warming Feed the Warming?* Oxford University Press, New York.

Taylor, D. R. and L. W. Aarssen. 1988. An interpretation of phenotypic plasticity in *Agropyron repens* (Gramineae). *Am. J. Bot.* **75**: 401–13.

Templeton, A. R. and D. A. Levin. 1979. Evolutionary consequences of seed pools. *Am. Nat.* **114**: 232–49.

Terborgh, J. 1985. The vertical component of plant species diversity in temperate and tropical forests. *Am. Nat.* **126**: 760–76.

Thoday, J. 1953. Components of fitness. *Symp. Soc. Exper. Biol.* **7**: 96–113.

Thomas, S. C. and F. A. Bazzaz. 1993. The genetic component in plant size hierarchies: norms of reaction to density in a *Polygonum* species. *Ecol. Monogr.* **63**: 231–49.

Thomas, S. C. and J. Weiner. 1989*a*. Growth, death and size distribution change in an *Impatiens pallida* population. *J. Ecol.* **77**: 524–36.

Thomas, S. C. and J. Weiner. 1989*b*. Including competitive asymmetry in measures of local interference in plant populations. *Oecologia* **80**: 349–55.

Thompson, J. N. 1984. Variation among individual seed masses in *Lomatium grayi* (Umbelliferae) under controlled conditions: magnitude and partitioning of the variance. *Ecology* **65**: 626–31.

Thompson, K. 1992. The functional ecology of seed banks. Pages 231–58 *in* M. Fenner, ed. *Seeds: The Ecology of Regeneration in Plant Communities*. CAB International, Wallingford, Oxon., UK.

Thompson, W. A., P. E. Kriedemann, and I. E. Craig. 1992*a*. Photosynthetic response to light and nutrients in sun-tolerant and shade-tolerant rainforest trees. I. Growth, leaf anatomy and nutrient content. *Aust. J. Plant Physiol.* **19**: 1–18.

Thompson, W. A., L.-K. Huang, and P. E. Kriedemann. 1992*b*. Photosynthetic response to light and nutrients in sun-tolerant and shade-tolerant rainforest trees. II. Leaf gas exchange and component processes of photosynthesis. *Aust. J. Plant Physiol.* **19**: 19–42.

Tilman, D. 1982. *Resource Competition and Community Structure*. Princeton University Press, Princeton, NJ.

Tilman, D. 1986. Evolution and differentiation in terrestrial plant communities: the importance of the soil resource: Light gradients. Pages 359–80 *in* J. Diamond and T. J. Case, eds. *Community Ecology*. Harper & Row, New York.

Tilman, D. 1987. Secondary succession and the pattern of plant dominance along experimental nitrogen gradients. *Ecol. Monogr.* **57**: 189–214.

Tilman, D. 1988. *Plant Strategies and the Dynamics and Structure of Plant Communities*. Princeton University Press, Princeton, NJ.

Tolley, L. C. and B. R. Strain. 1984. Effects of CO_2 enrichment and water stress on growth of *Liquidambar styraciflua* and *Pinus taeda* seedlings. *Can. J. Bot.* **62**: 2135–9.

Tramer, E. J. 1975. The regulation of plant species diversity on an early successional old-field. *Ecology* **56**: 905–14.

Tremmel, D. C. and F. A. Bazzaz. 1993. How neighbor canopy architecture affects target plant performance. *Ecology* **74**: 2114–24.

Tremmel, D. C. and F. A. Bazzaz. 1995. Plant architecture and allocation in different neighborhoods: implications for competitive success. *Ecology* **76**: 262–71.

Tremmel, D. C. and K. M. Peterson. 1983. Competitive subordination of a piedmont old field successional dominant by an introduced species. *Am. J. Bot.* **70**: 1125–32.

Turkington, R. 1983. Plasticity in growth and patterns of dry matter distribution of two genotypes of *Trifolium repens* grown in different environments of neighbors. *Can. J. Bot.* **61**: 2186–94.

Turkington, R., and J. L. Harper. 1979. The growth, distribution, and neighbour relationships of *Trifolium repens* in a permanent pasture. IV. Fine scale biotic differentiation. *J. Ecol.* **67**: 245–54.

Uhl, C., K. Clark, H. Clark, and P. Murphy. 1981. Early plant succession after cutting and burning in the upper Rio Negro region of the Amazon Basin. *J. Ecol.* **69**: 631–49.

Usher, M. B. 1987. Modeling successional processes in ecosystems. Pages 31–55 *in* A. J. Gray, M. J. Crawley, and P. J. Edwards, eds. *Colonization, Succession and Stability*. Blackwell Scientific Publications, Oxford.

Van Horne, B. and R. G. Ford. 1982. Niche breadth calculation based on discriminant analysis. *Ecology* **63**: 1172–4.

Van Hulst, R. 1979a. On the dynamics of vegetation: succession in model communities. *Vegetatio* **39**: 85–96.

Van Hulst, R. 1979b. On the dynamics of vegetation: Markov chains as models of succession. *Vegetatio* **40**: 3–14.

Van Hulst, R. 1980. Vegetation dynamics or ecosystem dynamics: dynamic sufficiency in succession theory. *Vegetatio* **43**: 147–51.

Van Valen, L. 1965. Morphological variation and width of ecological niche. *Am. Nat.* **99**: 377–88.

Vázquez-Yánes, C. and A. Orozco-Segovia. 1993. Patterns of seed longevity and germination in the tropical rainforest. *Annu. Rev. Ecol. Syst.* **24**: 69–87.

Via, S. and R. Lande. 1985. Genotype-environment interaction and the evolution of phenotypic plasticity. *Evolution* **39**: 505–22.

Vitousek, P. M. 1982. Nutrient cycling and nutrient use efficiency. *Am. Nat.* **119**: 553–72.

Vitousek, P. M. 1986. Biological invasions and ecosystem properties: can species make a difference? Pages 163–78 *in* H. A. Mooney and J. A. Drake, eds. *Ecology of Biological Invasions of North America and Hawaii*. Springer-Verlag, New York.

Vitousek, P. M. 1994. Beyond global warming: ecology and global change. *Ecology* **75**: 1861–76.

Vitousek, P. M. and J. S. Denslow. 1986. Nitrogen and phosphorus availability in treefall gaps of a lowland tropical rainforest. *J. Ecol.* **74**: 1167–78.

Vitousek, P. M. and P. A. Matson. 1985. Disturbance, nitrogen availability, and nitrogen losses in an intensively managed loblolly pine plantation. *Ecology* **66**: 1360–76.

Vitousek, P. M. and W. A. Reiners. 1975. Ecosystem succession and nutrient retention: A hypothesis. *BioScience* **25**: 376–81.

Vitousek, P. M. and L. R. Walker. 1987. Colonization, succession and resource availability: ecosystem-level interactions. Pages 207–23 *in* A. J. Gray, M. J. Crawley, and P. J. Edwards, eds. *Colonization, Succession and Stability*. Blackwell Scientific Publications, Oxford.

Vitousek, P. M. and L. R. Walker. 1989. Biological invasion by *Myrica faya* in Hawaii: plant demography, nitrogen fixation, ecosystem effects. *Ecol. Monogr.* **59**: 247–65.

Walker, L. R. 1991. Tree damage and recovery from Hurrican Hugo in Luquillo Experimental Forest, Puerto Rico. *Biotropica* **23**: 379–85.

Wallace, L. L. and E. L. Dunn. 1980. Comparative photosynthesis of three gap phase successional tree species. *Oecologia* **45**: 331–40.

Walters, M. B. and C. B. Field. 1987. Photosynthetic light acclimation in two rainforest *Piper* species with different ecological amplitudes. *Oecologia* **72**: 449–56.

Waring, R. H. and W. H. Schlessinger. 1985. *Forest Ecosystems*: Concepts and Management. Academic Press, Orlando, FL.

Waring, R. H., B. E. Law, M. L. Goulden, S. L. Bassow, R. W. McCreight, S. C. Wofsy and F. A. Bazzaz. 1995. Scaling tower estimates of photosynthesis with a constrained quantum-use efficiency model and remote sensing. *Plant, Cell and Environment* **18**: 1201–13.

Watt, A. S. 1947. Pattern and process in the plant community. *J. Ecol.* **35**: 1–22.

Wayne, P. M. and F. A. Bazzaz. 1991. Assessing diversity in plant communities: the importance of within-species variation. *Trends Ecol. Evol.* **6**: 400–4.

Wayne, P. M. and F. A. Bazzaz. 1993a. Birch seedling responses to daily time courses of light in experimental forest gaps and shadehouses. *Ecology* **74**: 1500–15.

Wayne, P. M. and F. A. Bazzaz. 1993b. Morning vs afternoon sun patches in experimental forest gaps: consequences of temporal incongruency of resources to birch regeneration. *Oecologia* **94**: 235–43.

Weaver, J. E. 1919. *The Ecological Relations of Roots*. Carnegie Institution of Washington Publication No. 286, Washinton, DC.

Weaver, J. E. and F. E. Clements. 1938. *Plant Ecology*, 2nd ed. McGraw-Hill, New York.

Webb III, T. W. 1987. The appearance and disappearance of major vegetational assemblages: long-term vegetational dynamics in eastern North America. *Vegetatio* **69**: 177–87.

Weiner, J. 1985. Size hierarchies in experimental populations of annual plants. *Ecology* **66**: 743–52.

Weiner, J. 1988. The influence of competition on plant reproduction. Pages 228–45 *in* J. Lovett-Doust and L. Lovett-Doust, eds. *Reproductive Ecology of Plants*: Patterns and Strategies. Oxford University Press, New York.

Weiner, J. 1990. Asymmetric competition in plant populations. *Trends Ecol. Evol.* **5**: 360–4.

Werner, P. 1979. Competition and coexistence of similar species. Pages 287–312 *in* O. T. Solbrig, S. Jain, G. B. Johnson, and P. H. Raven, eds. *Topics in Plant Population Biology*. Columbia University Press, New York.

Werner, P. A., I. K. Bradbury, and R. S. Gross. 1980. The biology of Canadian weeds. 45. *Solidago canadensis* L. *Can. J. Plant Sci.* **60**: 1393–409.

Whigham, D. F., I. Olmsted, E. C. Cano, and M. E. Harmon. 1991. The impact of Hurricane Gilbert on trees, litterfall, and woody debris in a dry tropical forest in the northeastern Yucatan Peninsula. *Biotropica* **23**: 434–41.

White, J. 1985. *Studies on Plant Demography. A Festschrift for John L. Harper.* Academic Press, London.

White, P. S. 1979. Pattern, process, and natural disturbance in vegetation. *Bot. Rev.* **45**: 229–99.

Whitmore, T. C. 1984. *Tropical Rain-Forests of the Far East.* Clarendon Press, Oxford.

Whitney, G. G. 1984. The reproductive biology of raspberries and plant-pollinator community structure. *Am. J. Bot.* **71**: 887–94.

Whittaker, R. H. 1972. Evolution and measurements of species diversity. *Taxon* **21**: 217–51.

Whittaker, R. H. 1975. *Communities and Ecosystems*, 2nd ed. Macmillan, New York.

Whittaker, R. H. and S. A. Levin (eds.). 1976. *Niche*: Theory and Application. Dowden, Hutchinson and Ross, Stroudsburg, PA.

Whittaker, R. H. and S. A. Levin. 1977. The role of mosaic phenomena in natural communities. *Theor. Pop. Biol.* **12**: 117–39.

Whittaker, R. H. and G. M. Woodwell. 1968. Dimension and production relations of trees and shrubs in the Brookhaven Forest, New York. *J. Ecol.* **56**: 1–25.

Wieland, N. K. and F. A. Bazzaz. 1975. Physiological ecology of three codominant successional annuals. *Ecology* **56**: 681–8.

Wiens, J., C. Crawford, and J. Gosz. 1985. Boundary dynamics: a conceptual framework for studying landscape ecosystems. *Oikos* **45**: 421–7.

Williamson, G. B. 1990. Allelopathy, Koch's postulates, and the neck riddle. Pages 143–62 *in* J. B. Grace and G. D. Tilman, eds. *Perspectives on Plant Competition.* Academic Press, San Diego, CA.

Willson, M. F. 1983. *Plant Reproductive Ecology.* John Wiley, New York.

Wilson, E. O. 1992. *The Diversity of Life.* Belknap Press, Cambridge, MA.

Wofsy, S. C., M. L. Goulden, J. W. Munger, S.-M. Fan, P. S. Bakwin, B. C. Daube, S. L. Bassow, and F. A. Bazzaz. 1993. Net exchange of CO_2 in a mid-latitude forest. *Science* **260**: 1314–17.

Woodrow, I., and K. Mott. 1988. Quantitative assessment of the degree to which ribulosebisphosphate carboxylase/oxygenase determines the steady-state rate of photosynthesis during sun-shade acclimation in *Helianthus annuus* L. *Aust. J. Plant Physiol.* **15**: 253–62.

Woodward, F. I. 1987. *Climate and Plant Distribution.* Cambridge University Press, Cambridge.

Woodward, F. I., G. B. Thompson, and I. F. McKee. 1991. The effects of elevated concentrations of carbon dioxide on individual plants, populations, communities and ecosystems. *Ann. Bot.* **67 (Suppl. 1)**: 23–38.

Woodwell, G. M. 1962. Effects of ionizing radiation on terrestrial ecosystems. *Science* **138**: 572–7.

Woodwell, G. M. 1963. The ecological effects of radiation. *Sci. Am.* **208**: 1–11.

Woodwell, G. M. and F. T. Mackenzie (eds.). 1995. *Biotic Feedbacks in the Global Climatic System. Will the Warming Feed the Warming?* Oxford University Press, New York.

Woodwell, G. M., R. H. Whittaker, W. A. Reiners, G. E. Likens, C. C. Delwiche, and D. B. Botkin. 1978. The biota and the world carbon budget. *Science* **199**: 141–6.

312 *References*

Wulff, R. D. 1986. Seed size variation in *Desmodium paniculatum*. 1. Factors affecting seed size. *J. Ecol.* **74**: 87–97.

Yodzis, P. 1989. *Introduction to Theoretical Ecology*. Harper & Row, New York.

Zangerl, A. R. and F. A. Bazzaz. 1983a. Responses of an early and a late successional species of *Polygonum* to variations in resource availability. *Oecologia* **56**: 397–404.

Zangerl, A. R. and F. A. Bazzaz. 1983b. Plasticity and genotypic variation in photosynthetic behavior of an early and a late successional species of *Polygonum*. *Oecologia* **57**: 270–3.

Zangerl, A. R. and F. A. Bazzaz. 1984a. Effects of short-term selection along environmental gradients on variation in populations of *Amaranthus retroflexus* and *Abutilon theophrasti*. *Ecology* **65**: 207–17.

Zangerl, A. R. and F. A. Bazzaz. 1984b. The response of plants to elevated CO_2. II. Competitive interactions between annual plants under varying light and nutrients. *Oecologia* **62**: 196–8.

Zangerl, A. R. and F. A. Bazzaz. 1984c. The response of plants to elevated CO_2. II. Competitive interactions between annual plants under varying light and nutrients. *Oecologia* **62**: 412–17.

Zangerl, A. R. and F. A. Bazzaz. 1984d. Niche partitioning between two phosphoglucoisomerase genotypes in *Amaranthus retroflexus*. *Ecology* **65**: 218–22.

Zangerl, A. R., S. T. A. Pickett, and F. A. Bazzaz. 1977. Some hypotheses on variation in plant populations and an experimental approach. *Biologist* **59**: 113–22.

Zwölfer, H. 1976. The goldenrod problem: Possibilities for a biological weed control project in Europe. *Plant Health Newsletter. EPPO Publ., ser. B* **81**: 9–18.

Index

Page references in **bold** denote illustrations